DISRUPTING SECONDARY STEM EDUCATION

This volume brings into focus the pivotal educational years during adolescence, when many learners are exposed to implicit and explicit messages that STEM is not a viable educational pathway for them.

Challenging this notion, *Disrupting Secondary STEM Education* brings together a collective of critical educators who share what disruptive STEM teaching looks and feels like from an insider perspective, as well as the ways they purposefully create curriculum to subvert existing structures that can confine learning. Through disruptive STEM teaching, a joy for learning is kindled, as well as a sense of empowered criticality in students that can support their development as global citizens facing complex futures. The collection shares stories across a spectrum of educators, from those beginning their teaching journey to those who've stood up against narrow curriculum and standardized testing for years in the capacity of both P-12 teachers and teacher educators. The voices of these educators illustrate how the work of disruptive STEM teaching can be actualized within cohorts of future teachers, achieved through early engagement with critical theories and generative field experiences that support and affirm a wide array of identities.

This book provides multiple theoretical and practical access points for the reader to understand the work of disruptive STEM teaching and offers a way forward for those interested in developing more critical curriculum in their own classrooms. As such, it will be important reading for postgraduate students and researchers in Social Justice Education and STEM Education, as well as for in-service educators.

Margery Gardner is the Director of Teacher Preparation and an Assistant Professor in Educational Studies at Colgate University, USA.

Critical Perspectives on Teaching and Teachers' Work

Series Editors: Nina Bascia, Denisha Jones, Arlo Kempf, and Rhiannon Maton

For more information about this series, please visit: https://www.routledge.com/Critical-Perspectives-on-Teaching-and-Teachers-Work/book-series/CPTTW

DISRUPTING SECONDARY STEM EDUCATION

Educator Experiences of Teaching for Globally Just Futures

Edited by Margery Gardner

Routledge
Taylor & Francis Group

NEW YORK AND LONDON

First published 2025
by Routledge
605 Third Avenue, New York, NY 10158

and by Routledge
4 Park Square, Milton Park, Abingdon, Oxon, OX14 4RN

Routledge is an imprint of the Taylor & Francis Group, an informa business

© 2025 selection and editorial matter, Margery Gardner; individual chapters, the contributors

Library of Congress Cataloging-in-Publication Data
Names: Gardner, Margery, editor.
Title: Disrupting secondary STEM education : educator experiences of teaching for globally just futures / edited by Margery Gardner.
Description: New York, NY : Routledge, 2025. | Series: Critical perspectives on teaching and teachers' work | Includes bibliographical references and index. | Summary: "This volume brings into focus the pivotal educational years during adolescence, when many learners are exposed to implicit and explicit messages that STEM is not a viable educational pathway for them. Challenging this notion, Disrupting Secondary STEM Education brings together a collective of critical educators, who share what disruptive STEM teaching looks and feels like from an insider perspective, as well as ways they purposefully create curriculum to subvert existing structures that can confine learning. Through disruptive STEM teaching, a joy for learning is kindled as well as a sense of empowered criticality in students that can support their development as global citizens facing complex futures. The collection shares stories across a spectrum of educators, from those beginning their teaching journey to those who've stood up against narrow curriculum and standardized testing for years in the capacity of both P-12 teachers and teacher educators. The voices of these educators illustrate how the work of disruptive STEM teaching can be actualized within cohorts of future teachers, achieved through early engagement with critical theories and generative field experiences that support and affirm a wide array of identities. This book provides multiple theoretical and practical access points for the reader to understand the work of disruptive STEM teaching and offers a way forward for those interested in developing more critical curriculum in their own classrooms. As such, it will be important reading for postgraduate students and researchers in Social Justice Education and STEM Education, as well as in-service educators"-- Provided by publisher.
Identifiers: LCCN 2024029315 (print) | LCCN 2024029316 (ebook) | ISBN 9781032498539 (hardback) | ISBN 9781032498522 (paperback) | ISBN 9781003395782 (ebook)
Subjects: LCSH: Science--Study and teaching--Social aspects. | Science--Study and teaching--Social aspects. | Science--Study and teaching (Secondary) | Mathematics--Study and teaching (Secondary) | Science--Study and teaching--Curricula. | Mathematics--Study and teaching--Curricula. | Critical pedagogy.
Classification: LCC Q182.8 .D56 2025 (print) | LCC Q182.8 (ebook) | DDC 507.1/2--dc23/eng20240816
LC record available at https://lccn.loc.gov/2024029315
LC ebook record available at https://lccn.loc.gov/2024029316

ISBN: 978-1-032-49853-9 (hbk)
ISBN: 978-1-032-49852-2 (pbk)
ISBN: 978-1-003-39578-2 (ebk)

DOI: 10.4324/9781003395782

Typeset in Galliard
by SPi Technologies India Pvt Ltd (Straive)

CONTENTS

ILLUSTRATIONS

Figures

Tables

PREFACE

Margery Gardner

This lifelong project begins with a connection to the land. I grew up in the foothills of the Adirondack Mountains on a small, economically struggling farm in a rural, predominantly White community. I spent my childhood summers discovering the insects, edible fruits, and mammalian creatures that co-inhabited our homestead. Summers especially were filled with adventures in the field and forestlands. As a child, I was always encouraged to explore my rural world with little to no adult guidance. My sisters and I would spend hours hiking access roads, looking out for toads, slugs, or other creatures to observe. My early love of learning was fostered through engagement with spaces and places mainly found outside the parameters of school.

Not understanding this at the time, I was developing an engaged science learning stance to view the environment in joyful and beautiful ways. Dopico and Garcia-Vazquez (2011) share how a connection to the land can act as a powerful learning experience for students, serving as a "positive pedagogy" that develops asset-based thinking around stewardship and sustainability (p. 311). As part of their project, students interviewed farming families and participated in tilling soil, protecting crops from predation, and caring for livestock. Students compiled and reported their findings, drawing tentative conclusions from their investigation. This type of learning centers on people with intergenerational knowledge of environmental stewardship. The power divisions between teacher and learner are complicated in productive ways by the added role of student as researcher. Dopico and Garcia-Vazquez's (2011) work illuminates the need to contextualize environmental science learning experiences outside the classroom, which directly resonates with my own formative ways of knowing. This book is one attempt to bring joy for learning into STEM classrooms in a critical and socially responsible way.

In 2009, I accepted a position as a secondary science teacher at a career and technical education school, and this is where I first began to tinker with dynamic learning spaces and curriculum. As part of a class on conservation, my students and I conducted a two-year-long aquatics exploration that first involved watershed mapping and water quality sampling of local streams. Students were responsible for tracking patterns associated with the variability of discharge rates and macroinvertebrate indices. Students also raised native trout species for eventual release as part of a greater cold-water conservation initiative. As a learning collective, we discovered first-hand the implications of environmental changes and human development on stream health and biodiversity. Students were able to see that their scientific work was not conducted in isolation and that human activities driven by capitalism inflict violence on our natural spaces. This teaching experience allowed me to view the possibilities of secondary-level education as an act of local environmentalism. My knowledge of the land felt validated and authentically part of an unbounded classroom.

I've spent nearly my entire adult life grappling with ways to teach science honestly and through a more critical and conscious lens while also working toward understanding myself as a White, cis-gender, settler educator. I'm constantly reconciling these normative identities and working toward understanding my situated positionality and the responsibilities that I assume both professionally and as a global citizen.

The research base for this book is rooted in my own experience as an informal educator, as a middle school and high school teacher in alternative public school settings, and more recently as a teacher preparer mainly in Upstate New York. I began the search for participants who might join their voices to this book in 2017 and since that time have interviewed several STEM teachers, both in-service and pre-service, and tracked their experiences across different times and locations. Through this inquiry, I've found community and inspiration, revealing countless stories of struggle, resistance, and liberatory education.

Authors of this book, in order of their contribution, include: Dr. Hugh Burnam (Hode'hnyahä:dye') is Mohawk, Wolf Clan from the Onondaga Nation, whose years of work in his field and commentary for Chapter 5 focus on indigenous education and conversations around the possibilities of decolonizing curriculum. Dr. Providence Rubey, who recently earned an EdD in the Mary Lou Fulton Teachers College in the Leadership and Innovation program at Arizona State University, discusses experiences around queering biological sciences. Enrique Nuñez, who teaches seventh and eighth grade math in the Southwest, shares experiences bringing social and political dimensions to his curriculum and instruction. Dr. Anita Bright, whose research is situated across intersectional boundaries to include multicultural/multilingual and math education. Dr. Randa Elbih, a social studies educator and former middle-grades teacher, looks to develop further supportive capacities for teacher candidates. Payal Patel, a teacher and teacher researcher, focuses on the nexus of mindfulness, learning, and mathematics. In solidarity with the STEM disruptors portrayed in this work, this authorship provides depth and symmetry to the potentiality of teaching. This research has been iterative and foundational to my own teaching of future educators and to my own identity. It feels both strange and familiar as it is written because of the intimacy I feel toward the data and subsequent analysis and the desire to share these experiences more broadly as they extend across time and context.

The structure of the book

The book begins with a review of the historical practices in STEM education that served to disenfranchise learners, especially Black, Indigenous, People of Color (BIPOC). The book then discusses the harms of tracking procedures in schools, especially at the secondary level, when identity formation as a STEM learner is most tenuous. The book leans into the construct of disruptive STEM education and describes its composition as critical, subversive, interdisciplinary, and situated by the self. The theoretical framing of disruptive STEM is actualized by educators in a complex formula based on the immediate needs of learners and the broader context. The book also brings into conversation the role of teacher education on the development of disruptive STEM teaching as pedagogy.

In the introductory chapters, the book features the experiences of four STEM disruptors, with different backgrounds with regard to race, gender, and geography. These STEM disruptors center on critical perspectives by finding openings to teach for justice, model advocacy for learners, and attend lovingly to their learners. In order to enact disruptive STEM, planning for instruction must intentionally resist methods that silo disciplines and position learners as passive participants. Chapter 2 discusses curricular designs that account for the unique outlooks of learners and offers space for expressions of collective humanity. In the second and third themes, the arc of the book then shifts in specificity to include author testimonials when engaging in disruptive STEM. The book offers examples of curriculum and first-hand reflective commentary that shed light on the experiences of educators adapting to a disruptive STEM theoretical framework.

The book demonstrates how teachers incorporate both content area knowledge and process-related skills to develop learner literacies and global citizenship. For instance, environmental racism in Flint, Michigan, discussed in Chapter 4, emerges as an egregious example of how disinvested communities must fight for access to clean water sources. Knowledge of corrosion, serial dilutions, and exponential growth are coalesced with conversations about systems of oppression in society. Process-oriented instruction poses compelling questions and inquiry-based activities that flex to allow burgeoning learner insights. Disruptive STEM education exposes systemic racism and settler colonialism as features that shape our current understanding of the world that necessitate dismantling and reimagination.

The book calls into question what sorts of content secondary-level STEM classrooms should cover, advocating for deep reflection and rethinking of the current curriculum. It positions secondary-level learners as adept and worthy of contextualized teaching with dynamic transactions occurring within the classroom across learners, teachers, and the community (Gardner & Tillotson, 2019). The anchors of love and hope can guide STEM education toward an anti-racist, decolonial future.

This book engages audiences such as education scholars and teacher preparers, as well as educators in the P-12 field, including administrators, inservice teachers, and professional development providers. Since the book focuses on STEM education for adolescents, it would be most compelling for educators who work within this content area and grade band. Members of educational advocacy organizations would also find this book applicable to their work.

References

Dopico, E., & Garcia-Vazquez, E. (2011). Leaving the classroom: A didactic framework for education in environmental sciences. *Cultural Studies of Science Education*, *6*(2), 311–326.

Gardner, M., & Tillotson, J. W. (2019). Interpreting integrated STEM: Sustaining pedagogical innovation within a public middle school context. *International Journal of Science and Mathematics Education*, *17*, 1283–1300.

ACKNOWLEDGEMENTS

I would like to acknowledge Dr. Micha Rahder for her caring guidance that shaped this work. I would also like to acknowledge Dr. Rebecca Mendelsohn and her team at the Longyear Museum of Anthropology for expertise and valuable insights. Thank you to Colgate University for support the publishing of this work.

THEME 1

Introducing disruptive STEM and basic approaches

INTRODUCTION

Setting the stage for disruptive secondary STEM education

Margery Gardner

Introduction

STEM education at the secondary level, in its existing form, fails to deliver adequate content and context regarding issues of global concern. Through a myopic view of STEM as comprised of siloed disciplines, learners are left to confront structural injustices such as racism, sexism, and heteronormativity outside the classroom. This chapter unpacks how these harmful social forces manifest in school practices, such as tracking based on perceived ability, influencing engagement in STEM and shaping long-term access to these subjects. Current teaching models in STEM education neglect the big picture need to develop global citizens. The central goal of this book, therefore, is to start a generative dialogue that reimagines STEM teaching practices in ways that actively resist historical legacies of harm and exclusion. While this book is rooted in the United States, these conversations transcend colonial geographies. Disruptive STEM teaching, as defined in this book, centers on critically examining the intersectional identities of both teacher and learner through pedagogical performances that are interdisciplinary and subversive.

Changes to secondary STEM educational practices is a matter of urgency. The devastating impacts of climate change and unsustainable resource use are felt globally in the form of long-term droughts, severe storms, and biodiversity losses. These "grand challenges" accelerate the need to thoroughly study STEM in schools in order to cultivate future generations that can work toward sustaining global ecologies (White et al., 2017). With the deleterious impact of humans as resource consumers (a colonial mindset),

DOI: 10.4324/9781003395782-2

STEM educators play an active part in shaping citizens who can make environmentally informed choices about water and soil resources, ecosystem services, and community climate resilience that work toward a healing orientation in service of our planet. As a collective, we can reframe related conversations toward collective goodwill, thus navigating "grand challenges" that extend beyond our lifetime and existing social imaginations.

Yet the values, beliefs, and cultural norms associated with STEM education are often not compatible with this "vision." This may be attributable in part to the fact that STEM teachers are imprinted with their own learning experiences in traditional schooling systems that reinforce normative thinking and compliance, inadvertently promoting models that are not well-suited to engage and include all voices. Undergirding these practices are school resources like textbooks and pre-packaged laboratory curricula, as well as structures that require adherence to overstuffed pacing guides (Aikenhead, Calabrese, & Chinn, 2006; Tytler, Osborne, Williams, Tytler, & Cripps Clark, 2008).

Looking at schooling across the spectrum, the transition from primary to secondary level should include a shift toward greater agency for students to make decisions about the content they study and how they learn. However, research suggests that the exact opposite tends to occur. Students at the primary grade levels experience STEM as a multi-modal process that includes hands-on engagement with concepts. However, once students transition to the secondary level, they are much more likely to be exposed to transmissive teaching styles. Tytler et al. (2008) describe transmissive teaching for subjects such as math, chemistry, and physics as a predictable cycling of the following: presentation of rules or principles, student practice, unfamiliar problems where students apply the rules/principles just covered, and finally the teacher providing the "correct" responses and evaluating student success. Under this pattern, knowledge is considered "received" from the teacher by the learner and is not open for contestation. STEM education is reduced to a superficial transactional interaction (Osborne, 2006).

Adolescent learners resent pedagogy that narrowly depicts STEM as linear and devoid of context. The results of transmissive approaches to STEM teaching include student apathy, anxiety, and boredom. The implications of such negative experiences, especially at the early secondary level, have staying power with students and can shape their life trajectories with regard to their future relationship to STEM (Tytler et al., 2008).

Mathematics, in particular, is a content area where learners quickly ascribe a failure identity to their abilities. Mathematics performance is under excessive scrutiny due to its high-stakes nature as a tested subject linked to teacher accountability in public schools in the United States and many other countries. Structures once again reinforce messages of "smartness" through ability tracking. There are clear racialized, sexed, and classed undertones regarding participation. This practice functions as part of a larger historical apparatus that denies oppressed people access to complete citizenship.

This book takes inspiration from the efforts of previous intersectional advocates who recognized the importance of STEM teaching and learning as a means to obtain full access to global citizenship.

Civil rights icon Robert (Bob) Moses developed the Algebra Project in 1982 to address systematic shortcomings in secondary math. "We are putting literacy, math literacy, on the table," he explains in his book *Radical Equations* (Moses & Cobb, 2002, p. 16). The Algebra Project supports Black youth from disinvested communities while elevating math literacies for all students. Algebra, as he asserts, is a tool necessary to operate sophisticated technologies through symbolic representation, and it is vital that students obtain competencies that empower them to speak this language fluently. Math is for all people, not just a select few who happen to attend P-12 schools that have robust supports and tax bases. The Algebra Project calls learners from all income brackets into challenging math classrooms rather than tracking them out at every turn. Moses positions mathematics education as a "tool for liberation" due to the relationship between math literacy and economic access. He refers to this act of teaching math literacy as a radical act because of its centrality in providing access to financial stability for all. Math literacy is a necessary condition of social change that can mend oppressions like the school to prison pipeline, which is all too familiar.

"Disrupting Secondary STEM Education: Educator Experiences of Teaching for Globally Just Futures" also leans on the foundational scholarship of Jeannie Oakes, whose work gave voice to the injustices associated with secondary STEM, especially with regard to the practice of tracking. Her work offers theoretical support for the argument that the current state of STEM education is, in fact, insufficient and disproportionately benefits White and affluent students. Adopting the lens of social movements as a political and historical force, much in the same tenor as Bob Moses, Oakes, and Lipton (2002) illustrate how community-based organizing can influence school reforms through a case analysis of one district that attempted to dismantle tracking policies. These policies had created a "two-school" structure, with White/Asian students in upper-level math courses and Black/Brown students in regular or remedial classes. The article describes how efforts to de-track these courses resulted in White, affluent parent pushback. These parents leveraged their political and social power to sustain the fabric of White supremacy in favor of their privileged children, with the expectation that their actions would protect their children's economic and political advantages later in life. The false notion of equal opportunity, for instance, in obtaining a higher education degree, is a manifestation of long-held meritocratic beliefs held closely in the United States. This book serves as a window to peer inside the work of teacher resistance within the confines of an imperfect system.

It is not by happenstance that higher-tracked classrooms predominantly contain White, middle to upper-class students, while lower-tracked classrooms are comprised of Brown and Black students. "Tracking helps create and legitimate a social hierarchy within a school based on perceived differences in student ability," Christensen writes (2000, p. 170). Oakes and Lipton (2002) trace this sentiment along political lines, arguing that the rhetoric of tracking reflects disingenuous sensibilities around equity that first gained traction in the United States with the No Child Left Behind Act of 2001. They assert that the precision of assessment required to appropriately sort students to meet their educational needs is simply not possible given existing structures. Instead, school decision makers rely on a collective imagination of a successful and teachable student rooted in racist, sexist, ableist, and classist perceptions. The subsequent damage on students' psyches as a result of low-ability labeling is crushing. The number of students of color that opt for STEM majors and persist within a White-dominated and culturally confining setting remains abysmally low. STEM educators are much too comfortable with tracking as a viable narrative in secondary schools.

There are some states in the United States that mandate additional services for math and English language arts (ELA) instruction for struggling students, referred to as Academic Intervention Services or AIS. This can be an opportunity to open the proverbial gate for students and expose them to alternative pedagogies that empower rather than constrict. Some students may enjoy this additional small group setting with more teacher oversight, while others may find it further stigmatizing and detached from their other learning prospects. Delpit (2012) mentions a teacher in Oakland, California, with many students labeled as needing special educational services, "but when she treated them as scholars, they behaved as scholars" (p. 45). In this classroom she described, instruction assumed academic and spiritual elements. Students were expected to choreograph dramatic performances and learn their own histories. These practices cultivated a sense of achievement and allowed students to view themselves in a more capable light. Delpit describes a reimagined space for supporting students' learning in STEM that incorporates an embodied experience that challenges their assumptions of self and STEM. "Disrupting Secondary STEM Education" mobilizes teachers to build on the strengths of the learners in their classroom and position them within the curricula and instruction fully as authors of their own educational journeys.

This book highlights the voices of teachers and educators that possess an array of intersectional identities based on race, gender, sexuality, class, ability, religion, geography, indigeneity, and language, among other differences. Each educator's journey and reconciliation of their intersectional identities as part of their disruptive efforts inform the ways that they interpret and enact disruptive STEM teaching. The authors demonstrate an interrogation of their own complicity (in some cases) as normative bodies, settlers, and White educators. Disruptive STEM teachers embark on an evolving process of critical identity development, listening, and unlearning. De-centering Whiteness and normativity is hugely important to actualizing disruptive STEM teaching. This book purposefully leans on an array of educators that identify in complex and obviously intersectional ways. This sends a collective signal to our learners that we all work in solidarity. The personal identity dynamics of the mixed authorship within this work are celebrated and used to ground commentary and analysis in honest ways.

What is disruptive STEM teaching?

Disruptive STEM teaching draws connections to the world in critical and compelling ways that center on social connections rather than positioning them as distractions from discipline-specific content. Disruptive STEM teaching involves the following four facets: (1) a critical stance that informs all subsequent practices, (2) subversion, (3) interdisciplinarity, and (4) involvement of situated self. This section unpacks these four facets, which together serve as a theoretical thread that will be drawn across chapters and different author experiences.

Criticality

This book centers critical theorists, including greats such as Paulo Freire and bell hooks, as well as more contemporary scholars such as Django Paris, in order to understand STEM teachers and teaching through a lens that views education as open, non-hierarchical, and worthy of collective nurturing. This book invests in a deep analysis of power within school-based systems and acknowledges how schools are emblematic of larger oppressive societal forces. Disruptive STEM teachers draw connections to the world in critical and compelling ways that center interconnectivity.

In particular, hooks' theorization of engaged pedagogy, as well as her conceptualization of the mind–body split in *Teaching to Transgress* (1994), acts as an academic compass for this work. hooks provides clarity about the prevailing attempts to sever students' intellectualism from their bodily interactions in the classroom. The mind–body split speaks to the ways that we limit students' bodily expressions of self, while simultaneously concealing our own. This book seeks to humanize the STEM teaching and learning process, where this tension is frequently palpable, but is rarely addressed. Chapter 9 explores this concept fully through author Payal Patel's experiences of using reflective meditation in secondary math classrooms as a means for students to see themselves as whole.

Extending the notion of engaged pedagogy more squarely in the arena of STEM, Ramesh and Patel (2013) explain the importance of praxis as part of critical STEM-focused pedagogy. "Praxis is related to thoughtful examination of taken-for-granted assumptions. Praxis refers to the steps people take to act on their emerging critical consciousness" (p. 99). Critical pedagogies challenge hegemonic views of systems and practices and empower non-dominant groups and identities to build coalitions to dismantle the structures that oppress them. STEM education requires preparation in skills, concepts, and processes. In a post-truth era, where STEM knowledge is openly politically contested, praxis-based pedagogy can support STEM literacies and further engagement in citizenship (Greenwood-Hau, 2021, p. 2). Learners that perceive science in narrow terms are also limited in their ability to examine social-political-economic factors that influence discovery and knowledge generation. This book repositions STEM education as part of a broader critical conversation that acknowledges past harms and histories, while contemplating radical futures. For instance, as explored in Chapter 3, the South American banana wars can be used to illustrate the interplay of politics, capitalism, and environment.

Disruptive STEM teachers push back on positivist knowledge systems and are open to understanding spiritual discovery while protecting and preserving sacred knowledge systems. Disruptive STEM teaching positions every individual in the room as worthy of STEM learning and as an active contributor. Disruptive STEM education is an ongoing process of critical identity development, unlearning, and acceptance of the teacher's knowledge as paramount. Disruptive STEM education is joyful and nourishing to all parties and accepting of varying thoughts, dreams, and social locations, with guided opportunities for students to advocate for changes necessary in their world.

A transformational shift in how we view the work of STEM teachers includes foregrounding students' heritage and cultural wealth and acknowledging culture as dynamic and continuously evolving (Paris & Alim, 2014). Embracing linguistic diversity within classrooms provides a means to forge sturdier bonds between learners, teachers, and curriculum. This argument builds on Paris and Alim's culturally sustaining pedagogical framework that calls for greater linguistic justice as a means to express understanding of the world and make sense of phenomena and interactions. In the United States, for instance, there is a long history of linguistic segregation, including bans on speaking Spanish and Indigenous languages. With these legacies still fresh, in order to combat these assimilation practices, teachers resist in ways that allow them to remain employed, yet recognize and affirm the identities of their learners.

Subversion

Subversion is described in the *Merriam-Webster Dictionary* as "a systematic attempt to overthrow or undermine a government or political system by persons working secretly from within." Subversion is important because, as the definition suggests, teachers may do this work in ways that are clandestine in many cases, even to their colleagues. The neoliberal expectations imposed on teachers at the secondary level are extensive, since this period of schooling is when the stressors of high-stakes standardized testing peaks. STEM disruptors subvert the "all-science" curriculum by finding windows through which to slip critical dimensions into conversation, offering students a glimpse into the fully dimensional realities of learning beyond the textbook status quo.

Disruptive STEM teaching involves purposeful acts of subversion that actively counter White supremacy. J. R. Givens' (2021) book, *Fugitive Pedagogy: Carter G. Woodson and the Art of Black Teaching*, describes subversive teaching as an act of Black resistance and a counter-educational project. He anchors this work conceptually in the historical example of Tessie McGee, a Black teacher in the 1930s, who taught at the only Black secondary school in Louisiana. The stakes for Black educators, whose professional fates were dominated by White school authorities, were exceptionally high during this time and include possible harassment by the Ku Klux Klan. At extreme personal risk, Tessie McGee told stories of Black history that included enslavement, rebellion, and societal contributions. Her actions were indicative of a broader movement by Black educators that traces its historical contours from enslavement, the Jim Crow era, and beyond to actively resist anti-Black literacies and shape learning in ways that actively disrupted the power structures held in place by and through White supremacy. Givens argues that the words selected for inclusion in Carter G. Woodson's texts, the pedagogical stance that Woodson advocated for, and McGee's subversive teaching operate in tandem as part of a greater historical legacy that "all embody a fugitive project." Such work was dependent on rich social networks of Black educators and scholars who could collaborate, given a common goal of Black educational liberation (Gibson, 2022). Rooted in these legacies of Black resistance, subversion in STEM education reframes non-confrontational content in order to render visible painful truths.

Subversion has been part of classroom practice since the existence of structured schooling. Yet, due to its more clandestine nature, subversive teaching and its impact on the field of critical education remain shrouded.

In order to demystify its inner workings, I refer to Postman and Weingartner's (1969) *Teaching as a Subversive Activity* initial framing. The authors refer to a "what's worth knowing curriculum" rather than teacher-proof scripted versions that are supposedly standards-based. Student interest as a key feature of formal schooling aligns with Dewey's earlier conceptualization. Dewey (1913) describes student interest as following: "Interest is not some one thing; it is a name for the fact that a course of action, an occupation, or pursuit absorbs the powers of an individual in a thoroughgoing way." *Teaching as a Subversive Activity* interrogates the possibility of radically restructuring schools to be fully grounded in student interests (p. 65). The authors advocate for the complete dismantling of existing curriculum and for its replacement with a version of schooling that allows students to generate their own worldly questions and determine their own educational goals, which would happen through inquiry-based journeys where teachers pose problems and allow space for student contemplation.

Postman and Weingartner (1969) assert that students should be taught in schools to critique culture, which they refer to as "crap detecting," and to think independently about complex subjects with minimal teacher interference (McDaniel, 2009). Postman and Weingartner's (1969) work aligns with that of equity advocates such as Jonathan Kozol, who also began writing critiques of school practices, in the late 1960s, that sought to expose the ills of education. They discuss the pervasive social issues of the 1960s, including political war mongering, environmental calamity, and police violence, all of which still plague us today.

After reading Postman and Weingartner's work, I admittedly felt disheartened that a book written well before I was even born could elicit such feelings of immediacy and was reminded of the educational struggles we continue to face in order to build new, more just worlds. Postman and Weingartner's (1969) work, as well as that of brave Black educators, establish a counternarrative to the standards-based rhetoric that continues to resonate. They urge us to think radically about education and to continue to press against cultural norms of practice that encourage complacency.

Subversive teaching can be defined as the contradictory act of fulfilling neoliberal obligations to follow national and state standards and structures within formal education settings *while* creating intentional opportunities to move students toward more just futures (Dyches, Sams, & Boyd, 2020). This adopted definition is germane when discussing facets of disruptive STEM education that are subversive in nature. More and more teachers turn to methods of subversion, given the current political reality of this work. For instance, in conservative states such as Florida in the United States, any course content that involves Black, queer, and/or feminist authors or narratives are under direct threat and removed or heavily censored. This sharp political terrain makes enacting subversive teaching a high-risk endeavor. Political aggression from all levels of government, hyper teacher accountability, and omnipresent student surveillance are cited as particularly thorny (Dyches et al., 2020).

Subversive teaching is context-specific and pushes up against disciplinary cultures and communities that are highly variable and dependent. Dyches et al. (2020) also address the need for subversive teaching that attends to disciplinarity, pointing out that subversive teaching is often overlooked or underwritten in critical educator circles. Resistance is inherent in subversive teaching because teaching itself becomes transformed into an act that opposes normative traditions. Disruptive STEM embraces resistance as part of the dialect of teaching as a mechanism to open up learning spaces to those often shut out or silenced.

Interdisciplinarity

STEM disciplines are conducted within complex social environments and therefore must be acknowledged as interdisciplinary work. "We do not need to create the whole: the whole already exists, we are in it…The part cannot know the whole; the part exists by virtue of the existence of the whole" (Young & Gehrke, 1993, p. 447). Scientific discoveries often involve the interaction and collaboration of many investigators, so are better illuminated through interdisciplinary STEM educational models (Grinnell, 2011). Interdisciplinary models elevate learning and allow for authentic engagement in STEM that are more reflective of actual STEM practices. Making the transition from novice to expert in a field requires opportunities to connect and apply knowledge from an area of study to new situations, and interdisciplinary learning models open entry points for this kind of knowledge transfer (Bransford, Brown, & Cocking, 2000). Disruptive STEM education mobilizes interdisciplinary models to shrug off existing silos that are part of the hegemonic structures of schooling and replace them with a broader and inclusive perspective (Venville, Wallace, Rennie, & Malone, 2002). For instance, understanding the historical timeline and subsequent ethical considerations that arose during the development of the smallpox or rabies vaccines are useful for learners to navigate the sorts of questions about immunization that arise today. Disruptive STEM acts as a lever to engage in uncomfortable and emerging understandings of the legacies of the past on current knowledge production while inviting in other worldviews. Literature, film, and art are embraced as partners in understanding STEM as part of global citizenship.

All STEM area subjects are not valued the same within the schooling structures of today (Gardner, 2017). Since ELA and math are tested subjects, other STEM subjects get scant airtime at the primary level. Science becomes distilled to non-fiction reading or tabled altogether. Technology coursework has undergone many transformations over the past decades. Industrial arts, the precursor to technology education (before the mid-1980s), included the physical manipulation of materials such as sawing wood to construct a birdhouse. Since then, the meaning of the term "technology" has expanded widely and linked discursively to efforts to "prepare students for the 21st century." Technology offerings now include computer science, including the manipulation of coding software, and engineering programs such as Project Lead the Way (NRC, 2014). Technology is often considered a fringe pursuit in the general education classroom, taught only after science and math are mastered (Tytler et al., 2008).

Engineering has gained more recognition with its inclusion in the United States–wide standards referred to as the Next Generation Science Standards. Bybee (2011) argues that science and engineering have many overlapping aspects and rallies behind this shift in thinking, writing:

> With the exception of their goals – science proposes questions about the natural world and proposes answers in the form of evidence-based explanations, and engineering identifies problems of human needs and aspirations and proposes solutions in the form of new products and processes – science and engineering practices are parallel and complementary.
>
> *(Bybee, 2011, p. 6)*

Interdisciplinary curriculum is most effective when interconnections are applied to areas at natural intersections. Students should be given the opportunity to interrogate complex problems early on in their secondary education that features both process and content learning activities. One possibility involves the introduction of engineering design projects. The process followed during engineering design projects includes an iterative framework of researching, problem definition, prototyping, iterating, and reflecting (Jordan, 2016). Through this process, failure is celebrated and learners' needs for tactile and peer collaboration are satisfied. Learners become more metacognitively aware and are afforded space and time to contemplate their own engagement and understanding that catalyzes agency. The interrelated nature of science and engineering is animated through pedagogies that locate content within a cyclical process of discovery.

Involvement of the situated self

We all carry with us our histories and identities that shape the ways we engage socially and academically. Much of the disconnect between students' lifeworlds or ways of being and STEM instruction is caused by a false notion of STEM as purely objective. Laboratory practices and the tools associated with STEM position it as distant from social sciences and the humanities. Beyond the content covered, disruptive STEM education therefore includes space for self and self-reflection. Identity stems, in part, from our lifeworlds, or the total sum of social interactions and reflections on our experiences (Heidegger, Macquarrie, & Robinson, 1962). The ways that we understand ourselves in relation to others should be considered an inextricable element of the process of teaching and learning. Discovery of the natural world requires acknowledgment of individual lifeworlds, which inform and construct our personal knowledge stores. Kozoll and Osborne (2004) support this assertion in the following: "If a union between science and the self is achieved, we can fully realize the potential science has to contribute toward this broader educative process" (p. 158).

Building on the idea of lifeworlds, the term "scienceworld" means the socially constructed setting where science is conducted. There is often a significant disconnect between students' lifeworlds and scienceworlds (Eger, 1992). In traditional settings, the tools of science teaching render learning unfamiliar and distinctly different from experiences outside of school. Like many educators, Bevilacqua and Enrico (1995) oppose the use of textbooks to support science learning, claiming that "they leave out extraordinary science, but also the science they deal with is not that normal" (p. 6). Textbooks do not include the historical nature of discovery, excluding multiple interpretations and neglecting to make the process of theory generation transparent. Disruptive STEM bridges the gap between students' lifeworlds and STEM worlds through teaching tools that are more recognizable and personally validating to students.

Without acknowledgment of the intersectional aspects of our identity, we cannot carry out honest discovery and knowledge generation. Brickhouse and Potter (2001) describe identity as "one's understanding of herself [*sic*] in relation to both her [*sic*] past and potential future. Identity refers to ways in which one participates in the world and the ways in which others interpret that participation" (p. 966). Nespor (1994) points out how the structure of STEM curricula and associated discourses used to convey content can impact learners' identity formation. For instance, they found that a physics classroom that focused heavily on standardized testing outcomes and discourses of rigor promoted a narrow physicist identity that was viewed as both unachievable and undesirable among students from non-dominant backgrounds. By non-dominant, I refer to those students who do not identify as White, heterosexual, Christian, non-disabled, or native English speakers, or who belong to a low-household income bracket (Sensoy & DiAngelo, 2017). In contrast, Bevilacqua and Enrico (1995) advocate for students to be able to use their own subjectivities to think about how knowledge is lodged within historical contexts. Building the capacity to understand scientific concepts in alternative ways is based in part on the recognition of lifeworlds. Scientists also engage in connecting science with their own lifeworlds.

STEM teaching has the potential to inform the self, support individual growth, and provide a means to dismantle structural oppressions that play out in our schools. Disruptive STEM education expands disciplinary knowledge beyond the borders of traditional subject silos. It offers a broadened view of teaching and learning that values a wide array of lifeworld experiences. Disruptive STEM education is one way to present a more unified view of human understanding, where lifeworlds are appreciated and baked into content area explorations in life-affirming ways. Rather than presenting a narrow bundle of content, students are exposed to content that is embedded in broader social issues that lack, at present, clear solutions. As a result, the role of teacher shifts from ultimate knower to liberator.

In summary, disruptive secondary STEM education involves criticality, subversion, interdisciplinarity, and situated positionality. These theoretical foundations will be threaded throughout the chapters in different ways, based on each author's experiences and worldly engagements. In each chapter, the author will alert the reader to theoretical connections that weave an underlying fabric across their experiences, discussing how they connect their work to the concept of disruptive STEM and its components.

Developing disruptive STEM capacities

Schools of education often require teacher candidates to take one social foundations course during their program of study. This course acts as the single opportunity to openly confront systems of oppression and their influence on knowledge production. This requirement is often viewed as tangential or unrelated to the "real" teaching of STEM by students and educators. Social foundations courses are portrayed as diametric to curriculum and instruction seminars in "pure" math and science, also referred to as methods courses that focus on decontextualized content in topical fashion. Some even equate the cultural foundations course requirement as much like "the math class the English majors put off taking until their senior year" (González, Moll, & Amanti, 2006, p. 215). The fissure between natural and social sciences is often perpetuated by teacher educators and the broader academy that falsely conveys knowledge as discrete and bounded.

This book argues that disruptive STEM teaching requires sustained exposure to critical pedagogies that allow opportunity to develop into an overarching stance. One course in social foundations is insufficient to pre-pare teacher candidates to resist systems that maintain White supremacy.

A culture shift in teacher preparation is needed to distribute social, politi-cal, economic, and historical milieus that shape our understanding of the world across all coursework in teacher preparation.

Through this more honest version of STEM, teacher candidates can viv-idly see the interconnectedness of education as an ever-evolving human endeavor. Teacher preparers shoulder the responsibility to expose critical and interdisciplinary theories, so that candidates can gain the necessary knowledge, skills, and self-confidence to engage in the work of disruptive STEM teaching even in their first days in the classroom.

The role of teacher education on the development of disruptive capacities will be interrogated further across multiple chapters in this book. Chapter 1 depicts the work of disruptive teaching through the lens of individual teach-ers who each recount the formative impact of their preparation experience. For some, teacher preparation offered a transformative window to assume pedagogical risks while still insular to neoliberal demands of teachers. Conversely, Shanaya recalls problematic aspects of her preparation that she actively resisted as a student to forge a more equitable pathway for future teachers. David in Chapter 3 discusses the learning process as a student teacher who leveraged disruptive pedagogy to center compelling questions of humanity. In Chapter 5, a teacher candidate poses the question, "How could White teachers talk about Haudenosaunee history and culture with predominantly White students without perpetuating whiteness either inten-tionally or inadvertently?" that results in decolonial musings by authors Hugh and Margery. In Chapter 8, three teacher educators use lesson study to reveal gaps and opportunities of their curriculum and unlock greater pos-sibility with each iteration. The work of disruptive STEM teaching is hum-bling and involves a concerted effort by teacher preparers to work in solidarity with candidates to confront hard histories and difficult realities. While teacher preparation is one conduit, there are many external forces also at play to feed into teachers' critical commitments, including, but not limited to, personal background, identities, geographies, talents, interests, and moral outlooks.

Teacher preparation programs could specifically foster capacities in navi-gating political censorship and teaching in contentious spaces. "Learning as an exciting, fulfilling, meaningful adventure actually gets in the way of accomplishing the objectives of classrooms driven by teacher-proof curricula, obsessive testing, and the fear of making or (in most urban schools) remaining on 'the list'" (Hatch, 2007, p. 311). Disruptive STEM is conducted within complex social environments that require scaffolded opportunities to bring into focus. Developing nimbleness to conduct meaningful engagements can be advanced through purposeful and steady mentorship.

Research and book details

This book is part of a larger qualitative inquiry to understand how schools internally operate and how STEM educators can shift the balance in their own classrooms to promote more ecologically aware, democratic citizens. Data for this book were gathered over a five-year period from 2017 to 2023 and in a variety of forms that included observations, formal and informal interviews, document analysis of curriculum, and reflective journaling. The time invested for theory development using qualitative methodologies such as narrative inquiry, portraiture, and ethnography/autoethnography is substantial, but the chapter authors were keen to pursue this research due to its pragmatic importance to program improvement as well as its contributions to the field. Throughout the book, the names of some participants have been changed to pseudonyms in an effort to protect confidentiality. Both I and other authors purposefully and frequently engaged with and interrogated our positionalities in the process of chapter writing.

The purpose of this book is to bring into focus the pivotal educational years during adolescence when learners are exposed to implicit – and, in many cases, explicit – messages about STEM education. Learners shut down, shut up, or are shut out of STEM, never to return. With the secondary school context as the hinge, this book unlocks the struggles and celebrations of an array of critical educators who act as disruptors of status quo, neoliberalized STEM teaching and learning. Chapters written by members of a collective of critical educators share what disruptive STEM teaching looks and feels like from an insider perspective and how they purposefully curate curriculum to subvert existing structures and learning experiences. In alignment with the series on critical perspectives on teaching and teachers' work, *Disruptive Secondary STEM Teaching* kindles a joy for learning and empowers criticality in students that can support them in their roles as global citizens facing uncertain futures. The book shares stories from across a spectrum of educators, from those just starting out their teaching journey to those who've stood up against narrow curriculum and standardized testing for years across P-16. Collectively, this project seeks to reconceptualize STEM education for secondary students through analysis and resistance of institutional patterns that continue to falsely frame and disenfranchise so many.

The book is organized into three major thematic sections that reveal the work of disruptive STEM teaching and the ways in which such pedagogies can be taken up in a variety of contexts: First, an introduction to disruptive STEM teaching; second, content-focused disruptive STEM curricula; and third, process-based disruptive STEM confessionals. Using specific in-depth examples and discussions, stories present ways that teachers develop and enact STEM critical pedagogies to inform political advocacy and to establish sites of strength and healing through STEM education. The book provides multiple theoretical and practical access points for the reader to understand the work of disruptive STEM teaching and offers a way forward for those interested in developing more critical curriculum in their own classrooms. The array of identities, backgrounds, and geographical locations of teachers and co-conspirators speaks to the durability of disruptive STEM teaching as a praxis.

Through portraiture, Chapter 1 explores the life and practices of four STEM teacher disruptors, each with different critical identities and outlooks, who enact this work across a variety of contexts. Chapter 2 introduces the work of curriculum planning for disruptive STEM teaching through a loving critique of known pedagogical frameworks that are remixed to create critical social learning opportunities. Chapter 3 illustrates how teachers can complicate science learning through political and economic contextualization using the example of monocultural banana farming, elucidating the threats of neo-capitalism on global ecosystems and sustainable native farming. Chapter 4 features another critical environmental case study, the water quality atrocity in Flint, Michigan, and describes introducing environmental racism to middle-grade students alongside core math and science competencies. In Chapter 5, Hugh Burnam, a member of the Wolf Clan from the Onondaga Nation, and Margery Gardner, a White science educator, discuss ways that STEM classrooms can amplify the artistry of Haudenosaunee Black ash basket weavers while locating intersections of cultural continuity and ecological health. Chapter 6, authored by biology teacher Providence Ryan, centers queer and BIPOC voices in science across multiple geographies. In Chapter 7, Enrique Nuñez, a middle-school math teacher, describes bringing political valuation of goods and services into classroom conversations, as well as the implications for the San Antonio community in which he is situated. Chapter 8 offers insights from a multi-year collaboration between three teacher educators in different fields –Randa Elbih in social studies education, Anita Bright in TESOL, and Margery Gardner in science education – who together sought to improve their pedagogy in service of emergent bilingual and multilingual learners through lesson study. Payal Patel shares reflective meditations she developed for secondary classrooms as a means to heal math-induced anxieties and reflects on their use in her own teaching in Chapter 9. Chapter 10, coauthored by Hugh Burnam and Margery Gardner, closes the book by contemplating STEM literacy and more inclusive citizenship futures.

References

Aikenhead, G., Calabrese, A., & Chinn, P. (2006). FORUM: Toward a politics of place-based science education. *Cultural Studies of Science Education, 1*(2), 403–416. https://doi.org/10.1007/s11422-006-9015-z

Bevilacqua, F., & Enrico, G. (1995). Hermeneutics and science education: The role of history of science, *Science & Education, 4*(2), 115–126. https://doi.org/10.1007/BF00486579

Bransford, J. D., Brown, A. L., & Cocking, R. R. (2000). *How people learn: Brain, mind, experience, and school.* National Academies Press. https://doi.org/10.17226/10067

Brickhouse, N. W., & Potter, J. T. (2001). Young women's scientific identity formation in an urban context. *Journal of Research in Science Teaching, 38,* 965–980. https://doi.org/10.1002/tea.1041

Bybee, B. (2011). Advancing STEM education: A 2020 vision. *Technology and Engineering Teacher,* 30–35. https://www.proquest.com/scholarly-journals/advancing-stem-education-2020-vision/docview/853062675/se-2?accountid=10207

Christensen, L. (2000). *Reading, writing, and rising up.* Rethinking Schools.

Delpit, L. D. (2012). *"Multiplication is for white people": Raising expectations for other people's children.* The New Press.

Dewey, J. (1913). *Interest and effort in education.* Riverside Press.

Dyches, J., Sams, B., & Boyd, A. S. (Eds.). (2020). *Acts of resistance: Subversive teaching in the English language arts classroom.* Myers Education Press.

Eger, M. (1992). Hermeneutics and science education: An introduction. *Science & Education, 1*(4), 337–348. https://doi.org/10.1007/BF00430961

Gardner, M. (2017). Understanding integrated STEM science instruction through experiences of teachers and students (Doctoral dissertation, Syracuse University). https://surface.syr.edu/etd/686

Gibson, L. (2022, March–April). Fugitive pedagogy: Jarvis Givens rediscovers the underground history of black schooling. Harvard Magazine. https://www.harvardmagazine.com/2022/03/features-fugitive-pedagogy-jarvis-givens

Givens, J. R. (2021). *Fugitive pedagogy: Carter G. Woodson and the art of black teaching.* Harvard University Press.

González, N., Moll, L. C., & Amanti, C. (Eds.). (2006). *Funds of knowledge: Theorizing practices in households, communities, and classrooms.* Routledge. https://doi.org/10.4324/9781410613462

Greenwood-Hau, J. (2021). Teaching facts or teaching thinking? The potential of hooks 'engaged pedagogy' for teaching politics in a 'post-truth' moment. *Teaching in Higher Education,* 1–18. https://doi.org/10.1080/13562517.2021.1965567

Grinnell, F. (2011). *Everyday practice of science: Where intuition and passion meet objectivity and logic.* Oxford University Press. https://doi.org/10.1093/acprof:oso/9780195064575.001.0001

Hatch, J. A. (2007). Learning as a subversive activity. *The Phi Delta Kappan, 89*(4), 310–311. https://doi.org/10.1177/003172170708900416

Heidegger, M., Macquarrie, J., & Robinson, E. (1962). *Being and time.* http://pdf-objects.com/files/Heidegger-Martin-Being-and-Time-trans.-Macquarrie-Robinson-Blackwell-1962.pdf

Hooks, B. (1994). *Teaching to transgress: education as the practice of freedom.* Routledge. https://doi.org/10.4324/9780203700280

Jordan, M. E. (2016). Teaching as designing: Preparing pre-service teachers for adaptive teaching. *Theory Into Practice*, *55*(3), 197–206. https://doi.org/10.1080/00405841.2016.1176812

Kozoll, R. H., & Osborne, M. D. (2004). Finding meaning in science: Lifeworld, identity, and self. *Science Education*, *88*(2), 157–181. https://doi.org/10.1002/sce.10108

McDaniel, T. R. (2009). Review of teaching as a subversive activity. *The Clearing House: A Journal of Educational Strategies, Issues and Ideas*, *82*(5), 249–250.

Moses, R., & Cobb, C. E. (2002). *Radical equations: Civil rights from Mississippi to the Algebra Project*. Beacon Press.

National Research Council. (2014). *STEM integration in K-12 education: Status, prospects, and an agenda for research*. National Academies Press. https://nap.nationalacademies.org/catalog/18612/stem-integration-in-k-12-education-status-prospects-and-an

Nespor, J. (1994). *Knowledge in motion: Space, time, and curriculum in undergraduate physics and management*. Knowledge, Identity, and School Life Series: 2. The Falmer Press.

Oakes, J., & Lipton, M. (2002). Struggling for educational equity in diverse communities: School reform as social movement. *Journal of Educational Change*, *3*(3), 383–406. https://doi.org/10.1023/A:1021225728762

Osborne, J. (2006). Towards a science education for all: The role of ideas, evidence and argument. *Paper presented at the ACER Research Conference: Boosting Science Learning – What will it take?*, from http://www.acer.edu.au/research_conferences/2006.html

Paris, D., & Alim, H. S. (2014). What are we seeking to sustain through culturally sustaining pedagogy? A loving critique forward. *Harvard Educational Review*, *84*(1), 85–100. https://www.povertyactionlab.org/sites/default/files/ParisAlim_What%20Are%20We%20Seeking%20to%20Sustain%20Through%20Culturally%20Sustaining%20Pedagogy.pdf

Postman, N., & Weingartner, C. (1969). *Teaching as a subversive activity: A no-holds-barred assault on outdated teaching methods-with dramatic and practical proposals on how education can be made relevant to today's world*. Delta.

Ramesh, M., & Patel, R. C. (2013). Critical pedagogy for constructing knowledge and process skills in science. *Journal Educationia Confab*, *2*(1), 98–105.

Sensoy, O., & DiAngelo, R. (2017). *Is everyone really equal?: An introduction to key concepts in social justice education*. Teachers College Press.

Tytler, R., Osborne, J., Williams, G., Tytler, K., & Cripps Clark, J. (2008). *Opening up pathways: Engagement in STEM across the primary-secondary school transition. Australian Department of Education, Employment and Workplace Relations*. https://deakinsteme.org/wp-content/uploads/2014/02/STEM_Opening-up-Pathways-July_08.pdf

Venville, G. J., Wallace, J., Rennie, L. J., & Malone, J. A. (2002). Curriculum integration: Eroding the high ground of science as a school subject? *Studies in Science Education*, *37*, 43–84. https://doi.org/10.1080/03057260208560177

Young, D., & Gehrke, N. (1993). Curriculum integration for transcendence: A critical review of recent books on curriculum for integration. *Curriculum Inquiry*, *23*(4), 445–454. https://doi.org/10.2307/1180069.

White, T., Wymore, A., Dere, A., Hoffman, A., Washburne, J., & Conklin, M. (2017). Integrated interdisciplinary science of the critical zone as a foundational curriculum for addressing issues of environmental sustainability. *Journal of Geoscience Education*, *65*(2), 136–145. https://doi.org/10.5408/16-171.1

1

PORTRAITS OF STEM DISRUPTORS

Margery Gardner

Introduction

This chapter shares renditions of disruptive STEM in the classroom as a broader movement to reimagine narrow educational practices. Classrooms act as a shared space that is shaped by both the teacher and learners through a combination of personal identities, interests, and outside commitments. This chapter focuses on teachers' critical work to satisfy the intellectual curiosities of their learners despite outside pressures such as anti-Blackness. Black and Wiliam (2010) refer to the classroom as a "black box," which is not well-understood from the outside and involves a complex set of considerations. This chapter discusses how these four secondary STEM teachers "make the inside work better" by assuming a disruptive stance that shapes both classroom environments and pedagogical decision-making in order to illuminate the politics that surround knowledge generation (Black & Wiliam, 2010, p. 82).

Disruptive STEM teaching operates within a guiding conceptual framework that considers critical, subversive, interdisciplinary, and self-reflective facets. Building on definitions of disruptive STEM from the introduction portraits of STEM disruptors extend these initial musings by exploring how personal background, cultural influences, and diverse geographies sculpt pedagogical enactment. This chapter explores how teacher identity and school context readily attract individuals to specific facets of disruptive STEM elements more strongly than others and shape their decision-making. For instance, Jeri explained how she resisted the uncritical curriculum imposed at the school level while also cultivating a love of literature

DOI: 10.4324/9781003395782-3

indicative of critical and interdisciplinary facets demonstrating criticality, subversion, and interdisciplinarity. This chapter offers readers compelling narratives as well as concrete tools for disruptive STEM implementation in the secondary education field through firsthand experiences.

A myriad of factors contributes to notions of teacher self, including relational discourses in the classroom as well as societal discourses. A teacher's identity is constructed, in part, through experiences in STEM as both a learner and a teacher. Gee (2000) defines identity as acting like a "kind of person" within a particular context and as situated across four different dimensions: biological factors, such as skin color and anatomy; institutional components like school structures; discourse identities, or language-based expressions of personality made both by an individual and about them, filtered through societal views; and affinities displayed through interests, like being outdoorsy or artsy. Through social analysis, we see that the first dimension, nature, is overrepresented in an effort to de-emphasize the power of institutional forces on shaping perceived identity status. An example of this dimension includes how people in poverty are often framed as requiring only personal interventions rather than an overhaul of an inadequate system. Institutional identities are defined as authorities and structures that result in a place or position within a social system. Discourse identities involve the construction of identity through language or how others actively describe individuals. Affinity identities are activities that individuals actively engage in, for instance, being outdoorsy or artsy, that also serve as descriptors. Hobbs (2013) contends that there is a close connection between teacher identity and teacher agency and further asserts that teachers should be encouraged to explore their ever-changing identities. Building on these understandings of identity, situated positionality stands as one of the core pillars of disruptive STEM teaching. The notion of situated self acknowledges that identity development is ever-evolving and can be either affirmed and upheld within schooling spaces or left malnourished without attention and care.

In order to peer within the "black box" of the disruptive STEM classroom, I settled on portraiture as the best methodology for contextualizing the work of STEM disruptors. Portraiture challenges the "researchers as all-knowing" paradigm by allowing the portraitist to co-construct knowledge alongside participants. While controversial due to its perceived lack of rigor and impartiality in the field of social science research, portraiture allows for the illustration of participant voice while also rendering visible the themes, relationships, and contexts that contribute to an aesthetic whole. By sharing moments of solidarity, this approach empowered the teachers profiled in this chapter as a celebration of practice (Dixson, Chapman, & Hill, 2005). I found portraiture as a methodology to offer greater accessibility to a wider array of readers while not diluting the intellectualism of the pursuit. "If we want to broaden the audience for our work, then we must begin to speak in a language that is understandable, not exclusive and esoteric … a language that encourages identification, provokes debate, and invites reflection and action" (Lawrence-Lightfoot, 2005, p. 9). Portraiture is emblematic of how we acknowledge the situatedness of our positionality and communicate our findings in educational resource more publicly through storying.

The numerous conceptual similarities between disruptive STEM and portraiture further justify this methodological exploration, specifically due to its embrace of critical perspectives, thoughtful expansion of disciplinary silos, and ability to write in the researcher alongside the participant. In addition, portraiture searches for goodness within participants in order to dismantle the deficit framing and pathologizing that frequently appear within educational research (Dixson et al., 2005). This search for goodness allows for participants and their work to be seen even when much of their daily practice may be subversive in nature due to institutional circumstances such as standardized curriculum and strict pacing guides.

The portraits depicted in this chapter emerge from a conglomerate of observations and conversations that speak to a set of firsthand experiences of STEM disruptors. All participants were relatively new to public school teaching, within their first five years. Participants were purposefully recruited based on their geographic location, range of experience, and intersectional identities. Some of these participants were acquaintances of mine prior to the start of the project, while others were not personally known until the start of the study. In the case of the latter, participants were recruited by word of mouth or through social media. The research project launched in 2017 and included a variety of research methods, such as formal and informal interviews, observations of practice, document analysis of curriculum, and reflective journaling. The interviews and school observations took place either in person (prioritized) or remotely, over several occasions. Interviews ranged from 30 min to an hour per session. Classroom observations typically included one or two lessons of roughly 45 min each per session. Pseudonyms were assigned at the onset of the study for each participant, and each participant was able to review this work prior to publishing to verify its authenticity.

This chapter highlights the individual work of each participant, all unified under a collective charge. First, Jeri shares how cultural work becomes part of disruptive STEM teaching, as well as critical literacy development. Second, Jason questions his identity as a "nice" White teacher and walks the reader through an afternoon at his school. D. troubles the notion of identity for his predominately White, rural students by recounting his connections to the environment growing up in the Global South. Shanaya locates self through a devoted approach to intersectional math and ethnic studies curriculum. Through centering humanity, these individuals provide insights into the negotiations, frustrations, and celebrations of disruptive STEM teaching as ongoing and evolving work.

Cultural work in a junior high biology classroom

Jeri is a Black, female-identifying teacher with two elementary-aged children of her own. She taught for several years in the south prior to seeking out a masters' degree in the Northeast and then once again returning to warmer geographies. This distance required that much of our interactions be remote, either phone interviews or observations using teleconferencing software. Jeri expresses a strong sense of pride when it comes to her own schooling experiences and mentions on several occasions how blessed she was to have many strong Black teacher mentors in her formative years. Whenever Jeri describes one of her most powerful mentors, her seventh-grade science teacher, she still seems very enamored. "He used to ride a motorcycle to school, he still worked at the hospital in the nighttime and taught during the day, he was super young, just cool, and I'm like, science looks like this?"

Given her strong proclivity for both the subject of biology and her admiration for her teacher, Jeri knew from an early age as a middle-school student that science would be part of her future career. A seventh-grade cow eye dissection lab solidified her choice. She entered a math and science-focused charter high school and continued to keep in touch with her seventh-grade teacher throughout her schooling. Her teacher would invite Jeri to different events both inside and outside school, demonstrating authentic and lasting care. Eventually becoming a building leader, Jeri's former teacher and lifelong mentor continued to stay in close contact with Jeri. He continued to steer her toward Historically Black Colleges and Universities (HBCUs) as she transitioned into higher education. "'You're a Spellman woman,' he would say." Jeri ended up studying biology at Howard University and was invited by her teacher mentor to speak to younger students about her educational journey. The modeling of care by her former teacher and ongoing mentor shaped many of Jeri's engagements with her own students, especially Black girls. She sat and talked with them during lunch periods, took them out to get their nails done, and even met up with friends, trekking across town with them on weekends.

In her first year at a new junior high-school science position, Jeri struggled to actualize her vision of disruptive STEM teaching. She attempted at many junctures to give students greater agency over their own education but found this perspective in direct conflict with the overall school culture of compliance-based curriculum and instruction. For her, the apathetic use of packet work by her colleagues ignited feelings of rebellion. In an exacerbated tone, she called out her fellow teachers during our conversations, "They do not want to do anything that requires more work!" Conversely, Jeri was constantly looking for social justice–related connections in her teaching and was undeterred by the labor involved in finding these materials. She says she can't get her former critical teacher preparations out of her mind.

> I pulled out my Rethinking Columbus Day book and there was mining for Uranium on native land and how it affects native people. I wanted to cover those things. It would have been great to do for Indigenous People's Day (Interview, 10/10/19).

She has many ideas that she'd like to put into practice but faces resistance from her all-White teacher counterparts who are perfectly content purchasing prefab worksheets from sites like Teachers Pay Teacher.

Jeri's school upholds a false and highly problematic stance that is anti-BIPOC (Black, Indigenous, and People of Color) and anti-queer to its core. Jeri entered the classroom primed to design the physical space to be as welcoming and desirable to non-White or non-binary students as possible. During one of her first acts as a teacher, just prior to the official start of the school year, she posted a rainbow safe space sign that was promptly removed. It was clear from the start that her politics were incompatible with the other school personnel. I could sense Jeri's immediate deflation after this occurred, which tempered the start of her first year of teaching postgraduate degree.

The amount of energy Jeri expended as a Black woman among uncritical White colleagues who had clearly never interrogated their own Whiteness was exhausting and unproductive. "Deepening understanding is not dependent on agreement, back and forth arguing inherent to winning or losing debates is not useful to this goal," explain Sensoy and DiAngelo (2017, p. 185). Jeri encountered racial aggression on a consistent basis and is one of the only staff members in her school who is willing to call out these harmful actions, even though the actions of her peers remained unchanged.

The students in Jeri's classroom in this suburban district possessed an array of racial identities. "It was the definition of a diverse school," as she put it. Jeri excelled in her first year teaching middle school science and quickly gained a reputation for exceptionalism and dedication. Even her problematic principal acknowledged Jeri's talent for connecting with students and supporting their progress as learners. In her second year, Jeri was asked by the administration to teach advanced science classes, which dramatically shifted the demographic composition of her students. Jeri, who had always dreamed of teaching kids who looked like her, now faced an all-White, predominately male student roster. Due to this dramatic shift in student composition, she found that her advanced biology content was now bundled with a further burden of explicit cultural work. She recounts a specific example:

> On Friday, I wore a dress, and I wore my hair wrapped up, and a kid in my advanced classes, [said] 'What do you call that thing on your head?' And I said, "a scarf, a hair scarf," and someone said, "a durag," and I said, "Oh, well, let's have this conversation…" I'm at a school where it's predominantly white and yeah, there's culture in that, but I feel like me being who I am, I come in with culture that they are not used to and so most of the times, I'm the, I'm the learning person. And so, we had a conversation: This is a durag. This is a headscarf, this is a bonnet, and even though it was a class full of boys, we had the conversation.
>
> *(Interview, 10/24/20)*

This experience reveals the cultural identity work ever present in teaching. Freire, Macedo, Koike, Oliveira, and Freire's (2018) *Teachers as Cultural Workers* describes identity as "a dynamic relationship between what we inherit and what we acquire" (p. 124). Freire depicts identity as something that is inherently individual yet couched within a hegemonic system that limits actualization for some. Our cultural and historical worlds collide within the social spaces of the classroom, where learning and teaching flow in multiple directions. Jeri's example acknowledges the cultural needs of students and for opening spaces for honest dialogue. This experience caused discomfort for Jeri who, when debriefing on the incident, said, "it's a blow to hear [from students]… 'What's that on your head?'" She also mentioned that it was a "class full of boys" after all, so the gendered connection with hair and hair styling clearly contained cultural boundaries that were made transparent.

Despite the personal vulnerability, especially considering the larger anti-Black sentiment in this school, she persisted in providing a window for understanding: "Okay, this is a scarf. This is a durag. This is a bonnet and have that conversation because it probably wouldn't have come up in any other space." She sensed in that moment the need for a mindful pause on presenting biological content in order to provide a glimpse into a cultural world that was unlike their own. In essence, she momentarily put the lesson goals on hold to subvert instruction and to prioritize humanity. Jeri recognizes that this opportunity for her students to understand Blackness in honest ways is rare. The explicit cultural work done on behalf of her students by

Jeri resulted in a short divergence from her intended lesson toward a more positive and complex construction of Blackness that challenges stereotypes.

Jeri expressed an acute understanding of her role as a cultural worker through her presence in particular spaces, as well as her decisions as a teacher. For her, she can extend this cultural work through a self-acquired collection of books for her students that feature texts with broad representation. Jeri explains how the idea to incorporate a lending library took form in her classroom.

> …I love to read, and because I teach science it's not really a thing to read books and science and so I thought, well, why not just have some books that I enjoy on the shelves for the students to read. And it started with *The Hate U Give*. I read it, what, two years ago, and I thought it was a great book that was my first young adult book I read as an adult. And so, I was like, oh, young adult books are great. I could read these. And so, it started with that book and then it just started to grow.
>
> *(Focus group, 10/24/20)*

Jeri noticed that students at her school had few opportunities to access contemporary texts with a social justice focus. One of her students lamented, "I don't want to read *To Kill a Mockingbird*, I stopped on chapter seven!" Jeri developed a lending system with library cards that students use to sign out their selections. In her first year teaching at her current location, she only had one copy of the library books in the collection and it caused bottlenecks for certain titles that were popular and desired by the students. Jeri elicited donations from family and friends and obtained some books at a reduced rate by carefully finding suppliers that charged less for educational materials. She was able to supplement her offerings to feature two books of each title by her second year of teaching. Jeri insists that students can borrow a book for as long as they need, and if students want to receive bonus points for reading, they have to sit down with her for an informal book chat. Jeri noticed a change in student behavior as a result of her lending library. When students have completed all of their science-related classwork, they gravitate toward the in-classroom library.

Jeri's collection included 158 books, with the majority of these (133) falling in the category of fiction; 22 are historical fiction, ten are science fiction, and the remainder are young adult or juvenile fiction. While mostly published in the last decade, Jeri also includes canonical texts such as Sandra Cisneros' 1984 piece *The House on Mango Street*, which tells the story of a 12-year-old Chicana girl growing up in Chicago in the Hispanic quarter. There were eight biographies and 12 non-fiction pieces in her collection at the time of analysis. An overwhelming number of books highlight protagonists from non-dominant backgrounds, including 90 books with prominent Black characters, nine with Latinx main characters, and two with Asian main characters. This curation highlights the need for educators to also shine a light on protagonists from other marginalized affiliations, especially Asian Americans, who have been recently targeted in the United States both by former President Trump and White supremacist groups since the onset of the COVID-19 pandemic. In my estimation, there are other marginalized identities that were also not included in this collection that a teacher beginning to choreograph their own set of texts may want to consider. The topic of indigeneity is only found in one of the historical fiction pieces. This limited offering could reinforce the dominant message that native people and voices exist in past tense rather than the contemporary. Voices from the disabled/crip and LGBTQIA+ communities could also enhance students' understanding of difference beyond racial identity parameters and lead to more intersectional conversations.

The number one theme that emerged from the collection is a spotlight on race, identity, and justice, with 43 books centering on this issue. Examples of these texts include one of Jeri's personal favorites, *The Hate U Give* by Angie Thomas, several books by Jason Reynolds, including *All American Boys*, and *This Is My America* by Kim Johnson. All three, among others, received widespread acclaim due to their ability to connect with readers and open up conversations about police brutality and racial violence. *All American Boys* was actually the third most banned book in the United States in 2020 due to its inclusion of anti-police messaging, alcohol, profanity, and drug usage. While visceral and unpleasant, these texts have a place in the schooling experience that disruptive STEM teachers can amplify.

This collection of books demonstrates how one teacher sees the world through literature and disrupts the idea that love of literacy and science are somehow incompatible. Paul Fleischman and Mechthild Hesse's (1997) *Seedfolks* is the only fictional text included that describes a community gardening initiative, seen through the lens of multiple participants. Three of the biographies included Margot Shetterly's (2018) *Hidden Figures*, William Kamkwamba's (2010) *The Boy Who Harnessed the Wind*, and Louis Haber's (1991) *Black Pioneers of Science and Intervention*. These offerings represent a diverse array of personal stories that weave in human ingenuity and persistence. Skloot's (2017), *The Immortal Life of Henrietta Lacks*, was also featured in Jeri's collection and serves as a powerful cautionary tale regarding ethical research and medical apartheid. A gap remained in offerings that highlight STEM counternarratives and unlock the beauty and power of indigenous knowledge.

"It's growing. It just takes a little bit of steam," reported Jeri. She had conversations with the librarian at the school and offered suggestions of high interest. Other teachers reach out to Jeri via social media for assistance during their own classroom library cultivation. Social media also served as a source of solidarity and inspiration, including from Instagram posts from White Girl Learning, Pernille Ripp, and We Need Diverse Books, to name a few.

Style (1996) introduced the concept of curricular mirrors that offer an opportunity for introspection and seeing one's truest self. BIPOC people have historically been denied access to literary mirrors. In 2021, the Centre for Literacy in Primary Education released a survey of ethnic representation in children's literature in England. In the prior year, only 8% of the children's books published featured a main character that identified as BIPOC. In the fiction genre, only 7% featured diverse ethnic characters. To put these statistics in perspective, 33% of children of primary school age are from a BIPOC background. Ironically, the study found that 33% of books published in 2020 included animals or non-human characters. This study verified what many educators have experienced firsthand. Representation of students from non-dominant, non-White backgrounds are dramatically missing. Through Jeri's work to bring books into her classroom that are Black-centric and grounded in justice, she once more embraced a disruptive stance that allowed her students to view themselves as powerful and complex in literary works.

Style (1996) also argued for the need to embed in curriculum various frames of reference that call into question our own realities and positioned knowledge. Style used the metaphor of windows that offer pathways into other selves and disciplinary ways of knowing. Jeri centered curriculum that offered such windows of understanding in her classroom. She was open to having honest conversations about race with her students in support of healthy exchanges. Junior high biology classrooms are not the most obvious setting for such conversations. Jeri located learning across natural and social spheres. There is often a vision of oneself that emerges on the glass surface of windows that showcases the material's reflective capacity as well. By complicating her role as a junior high-school biology teacher, Jeri passed along a critical ability to see oneself in relationship to others and the world.

"Nice" White Algebra teacher

Jason is a White, male math teacher who works at a public school in a large city in the Northeast in a predominately Latinx community. Like Jeri, Jason recently graduated from a liberal arts teacher preparation program with a critical focus. After he completed his graduate degree in secondary mathematics, he received additional certification in special education as part of a second master's degree. He speaks fluent Spanish, which is one reason he believes the school hired him. He is most likely the youngest of the STEM disruptive teachers highlighted in this chapter.

Since Jason was located on the eastern seaboard, I was able to visit his classroom on two separate occasions and conduct reflective in-person interviews, as well as phone conversations. Jason's school is a challenge to find due to heavy construction on the facade of the building. Just beyond the bank of doors to the school, there is a foyer that housed an entire security detail for the five different schools within this one building. The phenomenon of one building containing multiple schools is fairly common in the area, the premise being that reductions in student-to-teacher ratios better support individual student needs. The explosion of charter schools is due, in part, to this restructuring effort.

Jason explained that the school colors are maroon and black, but the color combination that demarcates the school looked bright red offset by black. A good portion of the students roaming the halls during passing wore black hoodies with the school's name in red. Jason explained that the sweatshirt was a vaccination incentive program at the school. The students milled about and offered friendly exchanges with the adults.

Jason's classroom is very square, with bulletin boards to the right when you enter and a panel of windows on the left. The bulletin boards alone bring back a certain aesthetic nostalgia. They have bright construction paper backgrounds with helpful phrases for "what to do if you get stuck or frustrated" tacked in place with a rainbow border. One bulletin board is dedicated exclusively to student work, with good grades from past assignments proudly displayed. Students line the room in long rows of paired desks. It seems as though the desk partners must have been self-selected: the overall energy in the classroom was palpable. Jason's intentional layout of the classroom space acts as a demonstration of what he expects and values. He celebrates the work of his learners and encourages them to persist independently when they struggle. Bob Moses discusses how math literacies are often portrayed as less important than other literacies, such as reading and writing (Moses & Cobb, 2002). Bob discusses how parents will commiserate with their child when they don't understand high school math, yet with this same child will read over their term papers to check for grammar and style. Jason, on the other hand, constructs an environment where students are pushed to do their best work and aim for comprehension of mathematical concepts. Partner work is normalized as a way to hold the learner accountable to the collective success of the class. This approach diverges from traditional notions of math instruction where independent seatwork is most prized.

Soon a group of high schoolers taking Algebra 2 occupy the desks and fill the space with chatter. This is the last period of the day, with a lively bunch of students occupying the desks, 18 in total. Posted on the board is a QR code for a Desmos site that contains lesson materials on average rate of change. Learners each grabbed an iPad and pulled up the lesson on the screen. The "Do Now" activity seemed straightforward and non-threatening, a question about exponential equations with a negatively correlated scatterplot. After a couple of minutes, they reviewed the Do Now question as a class. Jason held an iPad that allowed him to also screen-share as needed and track student performance in live time. Pedagogically, Jason's decision to select a low-risk initial question to start the class session allows learners of all levels of confidence to engage. The use of technology gives Jason a greater awareness of the needs of the learners in real-time formative data.

Jason then provided a table of values that contained a real-world example of credit cards, with the left column containing the number of months and the right column the balance. Credit card debit is something that many of us live with as a continued reminder of our consumer roles within a capitalist society. CNBC reported that the average American possesses over $6,000 in credit card debt (White, 2024, March 1). Fortune magazine reported that with inflation and pandemic-related economic strains for households in income obtainment, by the end of 2022 there may very well be levels of credit card debt larger than in the pre-pandemic era.

> During the last three months of 2021, credit card balances increased by $52 billion to $860 billion – the largest quarterly increase in the 22-year history of the data, according to the Federal Reserve Bank of New York's quarterly report on household debt and credit.
>
> *(Leonhardt, 2022, February 8)*

Connecting back to Bob Moses' argument about mathematics literacy as linked to financial stability, Jason's lesson pulls back the curtain on how capitalism functions to oppress large portions of the population, while only a relative few acquire more and more wealth. This lesson offers tools for learners to understand poverty as part as systemic inequity rather than just a personal circumstance, thus representing a way to enact critical pedagogy.

This example is representational of disruptive STEM practice because it directly addresses a societal problem and offers a space within the curriculum to explore the implications of personal debt through a critical, sociopolitical lens. Jason evokes critical pedagogy to address the pervasive social issue of personal debt by clearly mapping out, using mathematical visualizations, the need to question existing banking structures. This example takes actually minimal time during the class instructional period but reinforces a big picture message that mathematics classroom is a place to interrogate social inequities.

Understanding how debt works is part of informed citizenship. Since Jason teaches in a low-income community of color, this example may also provide a collective feeling of relief that someone is challenging the dominant narrative that asserts that financial burdens are not solely due to mismanagement of funds as the dominant narrative suggests. This example brings into focus the strategic nature of credit card companies to allow everyday people to shoulder immense personal debt that makes economic stability unattainable. Connecting this question to a larger unit on financial literacy could impact learners in powerful ways that would simultaneously shed the disciplinary boundaries between social science and mathematical instruction.

Once this conversation of the problem came to a natural transition point, Jason shifted the focus of the class to an entirely new context in order to achieve his overarching content-specific goal of understanding exponential equations. Jason seems to make a contextual leap by presenting another scenario to the class. While the flow of the lesson seemed to an outsider to be a bit jarring, upon reflection, Jason maneuvered in a way that allowed learners to engage in new ways with the same general concepts and offered a different perspective and point of common connection. He presented another problem that was relatable since it involved a common food source. Some learners mentioned that the example enlightened them about agricultural practices. Jason shared actual data obtained from a friend on her chickens' egg-laying habits. He shared a picture of the chickens, which incited unsolicited comments from the class. One student asked if the chickens eat pine cones. Jason replied, "I don't think so, it's just that the yard has pine cones in it." He had the class look up the formula and then calculate the rate of change over 7 days, given total eggs with a prompt as follows: On the first day of the week, Sara's chickens laid four eggs. Each morning she checks the chicken coop for new eggs and records the *total* number of eggs her chickens have laid so far this week (Table 1.1).

Jason then pressed for conceptual understanding, he asked, "What does it mean to get 2.83 eggs?" Students responded, "it's an average." Jason continued further, "Yes, it was an average rate of change. It's not quite 3. If my friend estimates 3, this is an overestimate." The lesson then returned to

TABLE 1.1 How many eggs? Lesson example

Day	Total eggs	Questions
1	4	1 Calculate the average rate of change for the week (from
2	8	Day 1 to Day 7).
3	11	2 A chicken can't lay part of an egg, so how can our answer
4	14	to question #1 means in the context of this situation.
5	17	3 Sara realizes she counted wrong on Day 7 and there
6	19	were actually 24 eggs. How would you change your
7	21	answer to #1?

its graphical start: "We're going to stick with the chicken data and use our data to make a scatterplot." Jason aptly navigated the virtual environment where student work is posted while making himself available physically by circulating about the classroom. Students seemed to drink up this one-on-one attention as a way to voice their math-related concerns without the entire class being privy to their conversation. He surveyed the classroom in a relaxed manner that was also deliberate in order to meet the specific needs of students. After some work time passed, the scatterplot was generated and Jason pulled it up on the screen in front of the classroom for collective review. He asked the class then, "How do you see 2.8 rate of change? How do you see what it looks like?"

About 30 min into the lesson, Jason presented a real-life complication to the scenario. He mentioned to the class that one of the tricky elements of the next problem is figuring out which information is useful for solving the problem and which numbers are merely distractors.

> Sara is sick of eating eggs, so she is going to start selling her chickens' eggs to her neighbors. She will sell eggs for $4.00 per dozen eggs. A dozen is 12 eggs. The chicken coop needs repairs that will cost $48. How many days will it take to pay for the repairs?

Jason encouraged the class to work together and provided scaffolding as needed for students to feel successful.

Jason embraces a constructivist teaching and learning approach that involves posing problems of emerging relevance. Jason allowed students the opportunity to familiarize themselves with the topic through an example that richly contextualized average rate of change. He artfully added extensions to problems that elicited critical thinking.

> Posing problems of emerging relevance and searching for windows into students' thinking form a particular frame of reference about the role of the teacher and about the teaching process. It cannot be included in a teacher's repertoire as an add-on. It must be a basic element of that repertoire.
>
> *(Brooks & Brooks, 1999, p. 44)*

As an exit ticket, there was one final question posed to the class that used a student's name as part of the example. "Rico is a TikTok influencer. He creates a table of values to record how many likes his newest dance video has each hour after he posts it." Rico was absolutely thrilled that he received a shout out during class and did a little dance move from his seat when the problem was first introduced. Jason mentioned after the lesson that Rico is the type of kid that "needs to be pulled in." Jason added that he frequently incorporates student names in math problems to make the experience more relatable. The disruptive STEM facet that resonates in this example is the involvement of situated self. Jason possesses acute knowledge of the learners in his classroom and finds ways to engage them best by activating their affinity identities such as TikTok-ing. As a relatively new teacher, Jason's practice continues to strengthen and build continuity across lessons and units for more unified exploration of critical content.

Teaching within the confines of system of standardized testing, Jason mentioned that he struggled at first to manage the demands of the classroom given the pressures of high school math to pass these tests. In Jason's first years of teaching, he reflected on his own persona in relationship to his students. Jason analyzed student behaviors to understand their needs, school-based or otherwise.

> I think I almost thought, maybe at first that, like, having a social justice stance towards teaching means that you're really nice and it definitely does not. I mean, it's high expectations which, for a lot of the students I teach, what they respond to best is being really strict sometimes. So, for me it sounds mean and for them it's just having high expectations.

Delpit advocates for an environment that provides support as well as academic challenge, especially for youth that have been historically marginalized. In her book, *Multiplication Is for White People* (2012), she described the term "warm demander" as the teaching ideal. A warm demander pushes students to do their best work and may not always be liked by their students. A warm demander acknowledges the cultural capital of the children in their class and leverages this to advance academic achievement. Warm demanders typically come from similar identity backgrounds of their students. They are able to admit when they have taken demands too far and be vulnerable in front of the class. In turn, students realize they are cared for and loved by their warm demander teacher and strive to meet the high expectations put forth. Unfortunately, the spirit of this concept is being taken up by hyper-neoliberal charter schools that use the idea of warm demanding as a tool for oppression. Teachers are expected to carry out a standardized set of expectations that fail to recognize the strengths but also the struggles of individual learners. Warm demanding can be a stand-in for practices that dehumanize students, such as calling out learners when they are unable to focus for long stretches of seatwork. Warm demanding becomes a way to justify placing blame on the learner for lack of educational success rather than the broader system that fails to support BIPOC learners. But Jason's interactions authentically reflect the ideals of the concept. Jason is keenly aware of his Whiteness in contrast to his classroom of learners and constantly modulates instruction to meet their dynamic needs. Most observable is his use of Spanish to reinforce directions to connect with individual learners, which he does effortlessly. Jason's outlook marks a contrast to Jeri's experiences where whiteness was both omnipresent yet never spoken about by White teachers. Challenges to White supremacist sensibilities came from Jeri directly, a Black woman who had to endure additional stressors in order to advocate for justice.

While debriefing on the lesson, Jason brought up the emotional labor of disruptive teaching and the pressures on teachers to support programs like credit recovery and test prep outside the confines of the regular school day. Jason commented that in the past he used to participate in many of the after schoolday activities, but by year three of the pandemic, he simply could not muster the energy to spend his Saturday commuting to the school. Over a cocktail in the hotel lobby where we met up, Jason shared how he prizes his summers and hopes to spend time with a new girlfriend who is a fellow math teacher. He expressed how taxing teaching has been for him, not just this year, and how there seems to be a magnifying effect as result of the COVID-19 pandemic that has led to outright exhaustion. Jason's experiences mirror other STEM disruptors and signal a need for a broader care movement for educators.

Challenging rural Whiteness

Care for self, land, and students emerged in D.'s testimonial as a STEM disruptor. D. is a middle-/high-school technology teacher in a rural, predominately White school district in the Northeast. D. identifies as a Black, heterosexual and male. At the time of the study, he had been teaching in this particular public school district for three years. We conducted our interviews remotely. He described his entry into the profession after pursuing several alternative career paths:

> I grew up in South America and moved to the US for college and eventually started a family here. A lot of my experience is all over the place and so with teaching, everything started coming together. I was an air traffic controller and that throws a lot of people. I was between the ages of 19 and 20 [squints eyes] before I started college so that was really my first real career. A lot of people would just stay in that career trajectory and retire from it but instead, I jumped into computer software and computer science so I have that as part of my background. While working [in a large city] I had to find a place to invest so I started some real estate. I quickly learned that hiring someone for renovations and repairs was almost prohibitively expensive so I started learning how to do this myself. Eventually I learned to do most of that construction and refinishing

myself. This brought me to my third skill set. My wife and I wanted to get out of the city because we didn't like what the lives of our young children were becoming even though on paper they had the lives that people were aspiring to, e.g., all the material possessions. However, we barely saw our kids. My wife and I both worked long hours to afford the stuff, but we didn't want a nanny to raise our kids anymore so we moved and started a farm. As we were starting to get our farm off the ground, we realized we needed supplemental income off the farm, in addition to health insurance. As we considered options, me becoming a teacher quickly became a front runner. Both my parents were teachers, my dad still teaches some courses in the Caribbean and my mom was a teacher for her entire career and retired as a principal. The household I grew up in had multiple teachers and, so, I'm comfortable with it and I like teaching in the most general sense. I like to find ways to explain concepts to varying audiences and teaching started to feel like a natural fit, so I started looking around. How do I start teaching?.

(Interview, 8/5/2020)

In some ways, D. also views teaching from the lens of cultural work, as indicated in the passage. Teaching felt like a natural pathway, especially given his exposure as a child to a family of teachers, where his many abilities could shine. He discussed his affinity for computer technology and agriculture, both relevant to his position as a STEM-based teacher. Throughout our conversation, D. positioned himself as a lifelong learner who was particularly intrigued with understanding function and design. Given the major career shifts D. made throughout adulthood as well as personal interests, D. makes evident that he values education as part of a broader project of spiritual and intellectual growth.

I tell my students don't think career. It's never a career, just think of your interests. After a while if you don't have the interest anymore, don't feel obligated to a thing or a place because they are not obligated to you (Interview, 8/5/2020).

D. explained how his disparate experiences enhanced his teaching of technology and its wide interpretations. At the time of the study, he was responsible for teaching small engines, computer science through Project Lead the Way (a curricular package that is largely problem-based), industrial arts-woodshop, foundations of technology, and a computer-based design class. "In terms of the teaching, I love exposing kids to random things. I try to guide them down whatever rabbit hole they want to go down." D. situates the self by placing deterministic power in the hands of the learner to find a career best suited to their abilities and interests.

Grounding his pedagogy in the needs and desires of learners took time and intentionality. He mentioned how a co-teacher in his classroom helped him to develop what he termed a "student-first mentality." He also attributes his successes, in part, to his teacher preparation program,

> The professors that I had [as part of his teacher preparation program] were also teachers in the high schools and they knew about the importance of relationships. I was able to lead with that and that really saved me my first year (Interview, 8/5/2020).

The growth he experienced by committing to understanding the learners in his classrooms helped him to gain insights into the community that he now was a part of.

Since D. was new to this particular community when he took the position, he had little exposure to the deficit framing of students based on familial lineage. D. strived to create a classroom community free of stigma. "I try to actively ignore what is being said about students unless they are in my classroom." D. made personal connections with individual students and found commonalities between his background and identity and his students' backgrounds and identities. "Hey, you're a poor kid from the country, I'm a poor kid from a country that's even poorer. I don't downplay their difficulty." D. made apparent the class similarities between his own upbringing and his students who come from struggling, disinvested townships. Class is something that can bind them, grounded in common experience. D. used class background as a pedagogical lever to build solidarity within the classroom that also created a starting point for race work.

Understanding how to do such race work is understudied in rural, predominately White spaces. Geographies play a substantive part in how race relations unfold in rural communities. Tieken's (2014) book, *Why Rural Schools Matter*, consisted of studies of two rural towns, Delight and Earle, in Arkansas, where the legacies of enslavement were felt differently among their citizens. In Delight, the landscape tends to be more wooded, and therefore, enslavement for cash cropping was not as much of a priority in the area. As a result, the community is more mixed, as made visible in school events such as basketball games, where there are Black and White people in the stands in nearly equal numbers. The school reflects these values with a Black principal and school board representation alongside a budding international program. Earle is comprised of rich delta land where enslaved Africans, later followed by poor Black sharecroppers, farmed to mass produce cotton for the benefit of White landowners. In this area, the Earle community was greatly divided along racial lines as these histories resonated throughout generations of its residents, often punctuated with episodes of violence against Black residents. Tieken work supports the argument that geography of place shapes the types of communities that take hold. Therefore, geography must be a consideration when developing STEM curriculum and instruction that seeks to disrupt White, middle-class sensibilities. There is a danger in drawing on place-less curriculum that fails to fold in locality and the nuanced cultural practices of a region. While Tieken suggests that these tensions are more palpable in the Southern United States, where White supremacy fuels flagrant acts to suppress Black people, I argue that in the Northeast, White supremacy isn't as muted as Tieken's scholarship may assert. Take a short drive across farm country or in the more mountainous areas and it will not take long to spot a "Fuck Joe Biden" sign, All Lives Matter/Blue Lives Matter banner, or a confederate flag. Such flagrant racist markers raise the stakes for disruptive STEM teachers, especially those who identify as BIPOC, making subversive practices most viable as a pedagogy for justice.

Rural communities in the Northeast have essentially erased racialized histories by asserting dominion as places for only White folks. For instance, in the area where I supervise student teachers as part of their teacher preparation program, there is a rich and engaging legacy of abolition and resistance that lies quietly buried among the rolling farm country. There were enslaved Africans and Black freed people who lived in these areas and farmed. In 1800, there were between 3 and 22 freed people in Madison County, according to the Historical Society, and at least nine enslaved Africans (Sernett, 2002, March 1). However, these legacies are barely visible, and detailed historical information seems to remain in the hands of local historians rather than embedded in the narratives of community and place. There is one exception: the small yellow and blue signs that pepper the land that designate places of historical significance. One such sign that subtly celebrates Frederick Douglass' presence in the area is just about the only identifier of this history to the general public (see Figure 1.1). I would wager that not many folks traversing the area take the time to read this sign or to contemplate its implications on community and identity. For D., who occupied a similar racialized landscape, I can easily understand his need to empathize with his predominately White students. He recognized the common struggle of working-class people as a critical juncture for dialogue. He approached race work delicately and with care for the learner.

D. is tall and well-dressed and stands in stark contrast with his White colleagues and the predominately all-White students who come from a farming community. When asked if the hypervisibility of his identity caused any personal discomfort, D. responded almost dismissively, describing this as

FIGURE 1.1 Fredrick Douglass state historic marker. Photo courtesy of Matthew Urtz.

"not unusual. This is my life. I've lived that space." He went on to explain the personal hardship of participating in the computer software industry earlier in his life. He noted that it was a frequent occurrence for him to be the only Black man in the room and for his contributions to be dismissed or downplayed by his White counterparts. "After a while, I stopped being apologetic for being a young black guy" in a competitive software engineering company leading cutting edge projects. "Navigating that is a part of my life."

D. described the feeling of being the only person of color in majority White school and community spaces as less adversarial than in past professional encounters. D. conceptualized Blackness as an amalgam that brings together his ethnic and geographic background, as well as the physicality of a racialized body. He explained, "There's an aggression that comes with owning your space and behavior." In the United States, Black Americans are expected to "navigate and survive," while D. had a different perspective growing up in South America with family ties to the Caribbean. "I grew up with black leaders, that's always been part of my experience. Non-White people took over leadership in many cases in the Caribbean while the US continues to suppress." In terms of background, he cites "my ethnic background. I wasn't born here and I didn't grow up in the States. Also, I lived in a large city and that seems to be intriguing to my students. All those pieces get their attention." D. complicates stereotypes that his students may possess and uphold by sharing with his classes his international upbringing and computer science affinities that are not normally associated with the Black experience in the United States. This creates cognitive dissonance on the part of the learner since they have to reconcile discourses of White supremacy that abound in the area with their teachers' perspective as Black man. Through demonstrations of mutual respect, D. is able to facilitate honest dialogue that helps students shed their preconceived notions of Blackness.

D. also brings similar care to the development of content area–specific goals. D. positioned care of the land as central to his teaching. It permeated all of the classes he taught, even those that seem less applicable. In this way, D.'s teaching draws from subversive disruptive practices through conversations of race, class, and environment even in unlikely courses such as small engine repair.

> In my classes, we talk about four wheelers and how they are a huge waste of resources without being disrespectful to a culture that values outdoor sports. These sports involve burning gasoline and other resources. We talk about being more resourceful with their use of cars. Planning trips rather than just jumping into cars. When they come up with the 'whys' The students pick up on my end goal and think ahead to ask what's the end goal? Why and in what ways is the environment impacted?
>
> *(Interview, 8/5/20)*

His love of the natural world sparked at an early age and was fostered throughout his adulthood.

> My father worked for the Organization of American States that is part of the United Nations in agriculture policy [in the Caribbean]. I grew up knowing that people are doing damage to the earth and the environment just for money.

D. described environmental advocacy as his teaching anchor and as "a huge part of what's driving my decision making." In this way, D. leverages several identity dimensions outlined by Gee (2000). His commitment to environmental stewardship is part of his affinity identity while assuming his institutional identity as a teacher to promote sustainable practices at natural intersections. Involving his situated self is part of an ongoing endeavor toward intersectional justice.

As a young professional, D. looked critically at his current lifestyle and wanted more connection to nature for himself and his growing family.

"It comes down to our kids' wellbeing and food," he states with clarity and conviction. Centering the environment and environmental sustainability is a continued part of D.'s curricular vision. Roberts and Green (2013) describe rurality as a concept that possesses demographic, geographic, and cultural dimensions. Tieken (2014) also reminds us that rurality is complex and includes the local tradition, cultures, and violence that have taken place in these geographies. Visiting D.'s farmstead, you instantly feel a sensory and spiritual connection with land. The scent of peonies permeates the landscape, animating visions of poultry sourcing small bugs and grubs in the greenery. D.'s vision of rural living is aesthetically enriching and life-sustaining. Rural does not equate to simple or quaint but can be a place of discomfort, growth, and renewal.

Finding devotion in ethnic studies

In contextual contrast, Shanaya's disruptive STEM work in math education is geographically distant from D.'s. Shanaya is a teacher-activist and scholar who I first heard speak to a class of introductory-level undergraduates. I was immediately impressed by how she addressed issues of justice with unapologetic fierceness. Her personal and teaching selves comingle in ways she described with a forthright confidence. While teaching in public schools for fewer than five years, she projects an air of total competence that seems to transcend this timeframe.

Our interactions were remote due to our physical locations. During our interview, Shanaya explained that her own schooling experience from both her P-12 years failed to validate her identity as an immigrant child of color. She mentioned that she was an "angry child" and that this emotion was fueled by the injustices she felt as a Brown kid in a mostly White schooling system. For instance, she wasn't selected to participate in the gifted program and her affinities toward math were not acknowledged as readily as her White counterparts'.

Shanaya's experiences in higher education as a teacher candidate unfortunately mirrored those in P-12 since her program neglected to affirm her identity and hyperfocused on the needs of White students. She recounts having to participate in a privilege walk activity that used her and the other students of color as "pawns to teach White people how lucky they are." She recalls one instance where a local Indigenous parent and a Mexican mother shared their experiences raising children in a large school system in the Northwest and were interrupted and subsequently silenced by a White professor. This event sparked a movement within the cohort to advocate for greater representation within the program. With the situation quickly escalating, Shanaya refused to complete the associated assignments required of the course. She was provided with an alternative writing assignment and ushered out of the program.

Shanaya mentioned that she actively sees racialization of her own math classrooms. "The stark differences were gross...can you guess which class looked like what?" she said with an air of sarcasm when describing her own classroom in her first year of teaching, where mostly BIPOC kids populated her section of "regular" geometry, while mostly White students were enrolled in an honors precalculus class. She added, "I teach mostly tenth grade. I should have had tenth graders across the board. I had a White class and I also noticed a difference, they were excited." The students understood math as a vehicle to unlock their future aspirations even if it didn't include mathematics as an end goal per se. In comparison, the other section of geometry embodied a very different affect. They didn't possess the enthusiasm to occupy the space positively. "They were burdened."

Shanaya went on to complicate this circumstance by adding that there were some Black students in the predominately White class and some White students in the predominantly Black class, as well as a host of other racial identities that were present in the space and that every student interacted with the materials differently.

> The energy was really different and that was on the system that created these energies. Not attitudes, because attitudes are a personal choice and energy is what people around you are giving off and feeding into. Energy is changing and moving. It's about the people and space (Interview, 7/6/22).

She adds that these energies are also shaped by teachers and how teachers project particular norms, typically White, middle-class norms, that either affirm or disaffirm individual students. Similarly, Yeh, Martinez, Rezvi, and Shirude (2021) authenticate the need for a systems-based historical lens when discussing the state of math education. "We are cautious here not to propagate myopic and pervasive understandings that sever mathematics education from the histories of colonialism" (p. 74). Shanaya reveals critical reflective capacities that support analysis of institutional structures. Student behavior is laden with bigger messages about identity and access. She advocates for greater opportunities for teachers to pause and reflect on their work in service to their students. Through this reflective work, Shanaya demonstrates disruptive STEM capacities as a critical teacher with an investment to explore the situatedness of identities within her classroom.

Shaped by what she witnessed occurring in tracked classrooms, Shanaya is motivated to develop a college-level ethnic studies math course at her school.

I want students to know what's happening politically around t because it helps students to understand and see that they are political pawns. And that they don't need to be and they have to stand up for themselves and they have to tell people the correct stories, right, because a huge part of Ethnic Studies is counter narratives.

(Interview, 7/6/22)

Shanaya's sociopolitical awareness directly informed her pedagogical decision to include counternarratives as part of her responsibilities as a disruptive STEM teacher. Ethnic studies acts as the framework in which her math teaching operates, with five guiding ethos to create a full and more loving version of teaching and learning. Yeh et al. (2021) name these ethos and illustrate how they function symbiotically. An ethos of identity, narratives, and agency is prioritized, especially in a political climate where White nationalist rhetoric has been gaining a stronghold. Locating counternarratives can reveal the power and intellect of BIPOC people. Counternarratives provide an interdisciplinary perspective on math content that traces histories and acknowledges cultural capital. An ethos of power and oppression makes visible the dynamics that undergird structural politics that actively seek to disenfranchise BIPOC people. This ethos confronts the notion of smartness and sheds light on how mathematical ability is often reduced to a narrow and Western-centric form that discounts different ways of knowing and doing math. An ethos of community and solidarity places students of color "out front" to provide for their needs as math learners (Yeh et al., 2021, p. 79). Community creation becomes an extension of learning and a means to draw strength and energy. The ethos of resistance and liberation strikes a balance between action and reflection, considering both as necessary for radical change in math education to occur that accounts for and celebrates the myriad of identities and voices frequently marginalized in school settings that don't neatly fit normative ideals. The final ethos celebrates intersectionality and the multiplicity of bodies, ideas, and histories, recognizing that intersectional structures impact each person differently and are greater than the sum of each racial, gendered, classed, and other identity present. Multiplicity refers to the duality of living in a Western-centric society as a non-dominant body. Multiplicity can act as a tool to uncover starting points where social change is possible and necessary, even if these are only small cracks in the massive foundation of White supremacy.

The ethos of ethnic studies mathematics seeks to identify fractures in the existing educational system where change can be possible, thus building on discussions of social movements by both Bob Moses and Jeannie Oakes from the introduction. Strong parallels can be drawn between the ethos of ethnic studies mathematics and the intentions of disruptive STEM education to bring about change by unifying secondary-level teachers through shared pedagogical commitments. There is inherent criticality involved in rendering visible hegemonic systems. Shanaya actively resists structural norms at her school and advocates on behalf of her students to offer a rigorous ethnic studies math course. She assumes a critical stance in order to push for greater opportunities for her students to be seen and heard. Through an interdisciplinary lens, ethnic studies mathematics animates the past, present, and future of communities of color as vital contributors to the field. Situated self is especially connective and opens up classrooms to the active use of counternarratives that celebrate historical mathematical contributions by communities of color.

The stakes for Shanaya are indeed high in relation to her advocacy for equitable ethnic studies programming within her school district. She was subject to death threats for her work in the development of an ethnic studies framework.

> I don't regret that because it's led to where I'm at today with ethnic studies, but it was a lot of, just, I spent a year doing a lot of unlearning, and a lot of reading and meeting people and having conversations and networking.
>
> *(Interview, 7/6/22)*

Shanaya recalled that it took three to four years to legitimize that class, and yet its viability as a future offering at the school was still shaky. The labor involved in the course creation left Shanaya feeling "tired of fighting." Like others working in solidarity to promote racial justice in their classrooms during an era of political conservativism, she discovered that support from administration was inconsistent and counterproductive. Shanaya opted to change schools for a fresh start with fewer preps as a designated precalculus teacher where the content may be limited. Shanaya doesn't seem deterred by the challenges of prescribed curriculum and seems eager for the opportunity to subvert it in ways that bring social-critical components to light.

Conclusion

The disruptive STEM teachers profiled in this chapter began their journey of development well before they ever stepped into a classroom. Jeri, Jason, D., and Shanaya all have strong impressions about their preparation and the ways it shaped their experiences once in the field. For some, teacher preparation functioned as space to contemplate radical teaching in a low-risk and theoretically rich environment with like-minded individuals. Their preparation program became, for some, a guidepost in which to internally check their curriculum for criticality. Formative coursework foregrounded the need for pedagogies of care and "student-first mentalities." Some filtered their experiences in higher education as fueling a Pollyanna version of social justice teaching that is unfettered by constraints of any kind and ironically simplistic in form. Others viscerally recount the limitations of their preparation program that caused major discomfort and awkwardness for not belonging to the dominant groups during exercises like a privilege walk that outed BIPOC students. In this case, negative interactions in teacher preparation actually catalyzed self-learning routines and deep reflection.

Shanaya, like Jason and many other teachers I've encountered recently, described recent career experiences as being tenuous and stressful. The stressful nature of teaching for justice, as demonstrated by Shanaya's lack of administrative support, echo in many ways Jeri's experiences with anti-Blackness in her first year teaching postgrad. Jeri ended up leaving her position to take a break from teaching but returned in a new capacity.

Gewertz (2019) found that BIPOC-identifying teachers cite examples like gestures of care for their BIPOC-identifying students that transcend well above their job descriptions. I think back to Jeri's example with her teacher who took her to extracurricular events and stayed in frequent communication well beyond her middle school years. Jeri, in turn, purchased books for her lending library out of her own pocket and taxied students around town to meet up to get their nails done on her days off. These actions leave mighty impressions on students but come at a cost for the teacher who must carve precious time away from family or work a second job to finance these extras. The system doesn't recognize this labor and offers little to no reimbursement.

Others cite limitations on their own curricular ownership and other administrative confinements. This seems to be the primary cause that Minnesota's 2020 teacher of year, Qorsho Hassan, left the classroom. Hassan is the first Somali American to ever receive this accolade. Her pedagogy stood out due to her incomparable ability to build rapport with children and teach difficult topics with honesty. She hoped to leverage her publicity to draw attention to an antiquated district practice, where those with the least seniority are the first to be removed from their positions if budget cuts arise. Hassan personally faced this issue at one point in her career and noticed the harm it caused other teachers of color just starting out in the field. The burdens outside the classroom imposed on Hassan and many like her are high. After the lengthy remote schooling periods brought about by the COVID-19 pandemic, students come back to physical classrooms with even more needs. Meanwhile, staffing shortages force teachers to take on many more responsibilities to make up for coverage issues, which takes away from planning and collaborative activities that sustain and enrich professional practice. With a wave of diversity, equity, and inclusion initiatives also hitting hard, often BIPOC-identifying teachers are tapped to take the lead in these well-intended but personally draining additions that again don't correspond with boosts in income. Hassan plans to take a year's leave that will most likely translate into permanent loss for second graders in Minnesota. Contending with constant strains of Whiteness compiled with pandemic extremes should *not* be on the shoulders of our BIPOC-identifying teachers who present a way forward (Dernbach, 2022, July 7).

In summary, each STEM disruptor described in this chapter spent considerable reflective work to understand the situatedness of their own identities in relation to classroom context and student composition. Furman and Traugh (2021) support this sentiment, recognizing that "teaching is highly situational," constantly challenging educators to leverage practical wisdom about their decision-making to "act in ways that are right, for children, for oneself, for larger educational purposes" (p. 15). These STEM disruptors lean on their personal funds of knowledge to create entry points to which content can adhere. For instance, Jason develops a problem set on the average rate of change that draws from the life experience of his friend who raises chickens. Jeri seeks to impart a love of reading, especially literature that centers Black joy, through the development of a lending library. D. helps students think more broadly about the contours of technology based on his varied past experiences and careers. Each teacher grounds their work in love and respect for their students that is demonstrated in ways dependent on geographies and personhood. For instance, Shanaya created a course offering that validates Latinx identities. Jeri consistently eats lunch with the Black girls in her class and often shuttles students around town for special events, even when she has substantial household obligations. D. cared for queer/trans youth by upholding their identities. STEM understandings are unlocked through contextualization that allows space to discuss issues of community and society more broadly. Subversive behaviors were prominent among all participants based on the limitations of curriculum, as well as school climates and leadership that upheld White supremacist goals. While each participant interprets their purpose as teachers through their own complex identities and knowledge structures, they each strive to improve current systems that track and quantify their students and replace these mechanics with dignity for all.

References

Black, P., & Wiliam, D. (2010). Inside the black box: Raising standards through classroom assessment. *Phi Delta Kappan*, *92*(1), 81–90. https://doi.org/10.1177/003172171009200119

Brooks, J. G., & Brooks, M. G. (1999). *In search of understanding: The case for constructivist classrooms*. ASCD.

Delpit, L. (2012). *Multiplication is for white people: raising expectations for other people's children*. New Press.

Dernbach, B. (2022, July 7). Minnesota's 2020 teacher of the year leaves the classroom. https://www.mprnews.org/story/2022/07/11/sahan-minnesotas-2020-teacher-of-the-year-leaves-the-classroom

Dixson, A. D., Chapman, T. K., & Hill, D. A. (2005). Research as an aesthetic process: Extending the portraiture methodology. *Qualitative Inquiry*, 11(1), 16–26. https://doi.org/10.1177/1077800404270836

Fleischman, P., & Hesse, M. (1997). *Seedfolks*. HarperCollins.

Freire, P., Macedo, D., Koike, D., Oliveira, A., & Freire, A. M. A. (2018). *Teachers as cultural workers: Letters to those who dare teach*. Routledge. https://doi.org/10.4324/9780429496974

Furman, C. E., & Traugh, C. E. (2021). *Descriptive inquiry in teacher practice: Cultivating practical wisdom to create democratic schools*. Teachers College Press.

Gee, J. P. (2000). Identity as an analytic lens for research in education. *Review of Research in Education*, 25, 99–125. https://doi.org/10.2307/1167322

Gewertz, C. (2019). Educators of color cite "invisible tax". *Education Week*. https://www.edweek.org/ew/articles/2019/10/02/educators-of-color-cite-invisibletax.html.

Haber, L. (1991). *Black pioneers of science and invention*. Houghton Mifflin Harcourt.

Hobbs, L. (2013). Teaching 'out-of-field' as a boundary-crossing event: Factors shaping teacher identity. *International journal of science and mathematics education*, 11, 271–297.

Kamkwamba, W. (2010). *The boy who harnessed the wind : Creating currents of electricity and hope*. Harper Perennial.

Lawrence-Lightfoot, S. (2005). Reflections on portraiture: A dialogue between art and science. *Qualitative Inquiry*, 11(1), 3–15. https://doi.org/10.1177/1077800404270955

Leonhardt, M. (2022, February). Americans are bingeing on credit card debt. It just grew at the fastest rate in 22 years. *Fortune*. https://fortune.com/2022/02/08/credit-card-debt-fastest-growth-22-years-holiday-shopping-new-york-fed/

Moses, R., & Cobb, C. E. (2002). *Radical equations: Civil rights from Mississippi to the Algebra Project*. Beacon Press.

Roberts, P., & Green, B. (2013). Researching rural places: On social justice and rural education. *Qualitative Inquiry*, 19(10), 765–774. https://doi.org/10.1177/1077800413503795

Sensoy, O., & DiAngelo, R. (2017). *Is everyone really equal?: An introduction to key concepts in social justice education*. Teachers College Press.

Sernett, M. (2002, March 1). *Petersboro national abolition Hall of Fame and Museum*. New York Almanack. https://www.newyorkalmanack.com/2022/03/peterboros-national-abolition-hall-of-fame-and-museum/

Shetterly, M. L. (2018). *Hidden Figures*. HarperCollins.

Skloot, R. (2017). *The immortal life of Henrietta Lacks*. Broadway Paperbacks.

Style, E. (1996). Curriculum as window and mirror. *Social Science Record*, 33(2), 21–28.

Tieken, M. C. (2014). *Why rural schools matter*. UNC Press Books.

White, Alexandra (2024, May 1). *Average American credit card debt by state*. CNBC. https://www.cnbc.com/select/average-credit-card-balance-by-state/

Yeh, C., Martinez, R., Rezvi, S., & Shirude, S. (2021). Radical love as praxis: Ethnic studies and teaching mathematics for collective liberation. *Journal of Urban Mathematics Education*, 14(1), 71–95. https://doi.org/10.21423/jume-v14i1a418

2

PLANNING FOR DISRUPTIVE STEM

Margery Gardner

Introduction

Teachers are caught in a complex web of responsibilities and situations that you could never anticipate until in the field. The barrage of canned curricular materials funneled to in-service teachers serves as fuel for neoliberal agendas. To understand the implications of using such materials and showing restraint when needed to foster holistic student learning are acts of both resistance and love. This chapter discusses how these dynamics inform the curricular planning process for disruptive STEM teaching. As a White teacher, I hope to continue to do the reflective work of understanding self and spaces in order to provide curricular opportunities that offer for learners to do the same. Whiteness as a conduit for White supremacy prevails when we look too narrowly at pedagogical possibility and practice.

The connectedness between planning, instruction, and assessment is undeniable. Central to curricular planning for disruptive STEM is bringing visibility to the social implications of science and technology and their influence on our learning, ways of expression, and personal freedom. Without urgent attention to relevance, the purpose of teaching and learning deflate. Centering questioning above formulaic objectives places agency with students as a process-based endeavor toward understanding that generates further inquiry. The versatility of question creation and posing is infinite. Bringing the textures of the world into focus through critical, interdisciplinary, self-guided questions is an act of justice in its fullest sense.

DOI: 10.4324/9781003395782-4

As Delpit articulates in (2012) *Multiplication Is for White People*, all students benefit from high academic expectations and access to opportunities that stimulate critical thought and social analysis. This chapter highlights the curricular development side of disruptive STEM teaching. Drawing from my work on teacher preparation, it offers planning perspectives that can be leveraged to develop criticality within learners' minds. The chapter offers a "loving critique" (a term coined by Paris and Alim) of Understanding by Design (UbD), a curricular framework that has been canonized as best practice across all disciplines in education.

This chapter seeks to interrogate existing planning structures like UbD while contemplating other more adaptive models. As an example of the latter, it examines the enactment of plans with core ethical questions about vaccine development in a rural biology classroom in Upstate New York, where vaccination hesitancy remains a pervasive issue. Critical media literacy can be fostered through disruptive STEM engagements as an imperative with the onset of fake news campaigns that seek to illegitimate science. Adaptive models of planning accommodate emerging technologies that can promote inclusive teaching. Educators also must reconcile with the role of technology in the classroom, including artificial intelligence (AI), as a new planning reality emerges. AI from a systems-level vantage point is also discussed as a tool of surveillance. Disruptive STEM teachers and learners must stay abreast of the uses of AI in schools and their potential to racially discriminate.

Understanding STEM education by design

UbD is a highly lauded and widely celebrated approach to lesson planning that signifies greater intentionality and alignment when generating lesson goals, assessment, and learning activities. Through the creation of the UbD framework, Wiggins, Wiggins, and McTighe (2005) set into motion a way for teachers to support their students as they explore big picture ideas that can challenge prior assumptions and convey material as confounding and at times unanswerable. They assert that curricular work must start, somewhat counterintuitively, with desired outcomes and key understandings. This approach is considered a results-based design rather than a content-based design.

The two approaches diverge in an important way. A content-based design assumes that students will learn through the appropriate selection of topic, text, learning activity, and assessment without accounting for the underlying meaning of the topic and text and how they relate to the student and their perspectives. Content-based design upholds traditional teaching methods normalized through historical schooling practices. It sets up learning opportunities that have potential to be meaningful for the learner but limit transparency for the rationale behind the topic, text, activity, and assessment components. Learning could happen, but without theoretical supports it most likely falls short of resonating with students over the long term. A results-based approach, in contrast, provides students with curricular opportunities to grapple with broad and meaningful ideas, such as the dangers of a single story or how capitalism rules the world. A results-based approach looks for the spark or seed that can perplex and puzzle students and moves toward greater learner agency.

The teacher candidates I work with become familiar with results-based approaches for planning after their introductory coursework is complete. They are asked to develop lessons with three primary components that mirror UbD's stages: (1) identify desired results, (2) determine acceptable evidence, and (3) plan learning experiences and instruction (Wiggins, Wiggins & McTighe, 2005, p. 18). Two program completers shared that their planning evolution included a marked mental shift away from an initial focus on learning activities. "I originally approached it as a fun connection activity. I was focused on engagement because of my experience teaching in 8th grade," one commented. Bringing intentionality to lesson planning became a greater priority, both horizontally by developing cohesive units and vertically by drawing from funds of knowledge and other classes. These teachers found that the labor involved in planning, while sizable, resulted in lessons where "the students got more out of it."

Central to the backward design model is posing essential questions that frame unit objectives. "The questions thus serve as doorways or lenses through which learners can better see and explore the key concepts, themes, theories, issues, and problems that reside in the content" (McTighe & Wiggins, 2013, p. 5). Questions act as provocations that give students the access points to see the utility of understanding the content alongside oneself. Essential questions are written in student-friendly language and embrace the human struggle to understand the world around us in functional ways. Mathematics-based examples of an essential question posed by the author include: Is there a pattern? How does how we measure influence what we measure (or don't measure)? In science, essential questions may appear as follows: How do we decide what to believe about a scientific claim? How do we measure what we cannot directly see? Why and how do scientific theories change? (McTighe & Wiggins, 2013, p. 4). One program completer elaborates, "The essential question is very important and how you frame it. If you don't frame it in a way that is clear they are confused."

The creation of backward design came from a place of resistance but has been co-opted to some extent by neoliberal accountability forces.

> Backward design was conceived as a process to denaturalize educational aims that had become overly rigid – for example, in cases where specific course content or instructional activities became so ingrained in an instructor's thinking as to become educational ends in themselves.
>
> *(McCreary, 2022, p. 85)*

Association for Supervision and Curricular Development (ASCD) leverages the success of UbD and placed a strong emphasis on UbD during the shift to Common Core State Standards in 2013, when many teachers were scrambling in the United States to develop new curriculum. The use of learning objectives as part of the UbD framework has been distorted to become yet another method of oversight; evaluating student and teacher performance rather than promoting an internal classroom compass to gauge learning.

Secondary-level teaching comes with plenty of bureaucratic burdens that splinter planning time and offer little space for honest reflection. In my last year of public school teaching, they reduced the number of science teachers at the career and technical school by half and I suddenly had hundreds of students on my roster. The weight of simply entering attendance and grading was immense. The struggle to manage the professional responsibilities of teaching while staying true to oneself and one's students is an ongoing struggle. UbD offers comfortable common language for planning but can push out alternative thinking, the opposite of what was sought during its inception.

UbD is only one of many guiding framework curricular perspectives that can be entertained while planning for disruptive STEM engagements. McCreary (2022) suggests a shift away from pedagogies of curing that insist educators impose a prescribed learning outcome to curricular work. In replacement, pedagogies of coping see students for who they currently are and what they readily recognize. More radically, a pedagogy of non-striving creates a classroom that is responsive to the most pressing needs of the learner, including making sense of political polarization or grieving personal tragedies. Strict enforcement of curricular frameworks such as UbD clouds the abilities of educators to think in unbounded ways and to actualize methods that enact such liberation from standard objectives.

McCreary (2022) offers pertinent considerations when developing curriculum that recalibrates based on the learner as a whole person, beyond learning outcomes. First, learning is intrinsic and cannot be separated from the learner themselves. Therefore, anticipating narrow goals and objectives for a lesson is unnecessary and hampers learning as an exploratory and intimately personal project. Second, providing adequate space in the curriculum for spontaneity is essential. Rather than a caveat that lightly gestures toward lesson flexibility, building in variability as "part of the design" can allow the learner to grasp content in ways that offer the most meaning. Pendleton-Jullian and Brown (2018) refer to this as "designing for emergence," noting that "the conventional machinations of instructional design can fall short of addressing our holistic needs" (p. 79). Emergence becomes more than a contingency plan but rather a reimagination of what the learning landscape should value. Expressions of emotion as students confront racism, bigotry, or violence are not typically included in the lesson planning process, yet teachers should offer time within instructional periods for learners to process after traumatic events.

Postman and Weingartner's (1969) work on subversive teaching radically proposes a thought experiment where all curricula are removed from the classroom and learning is conducted solely through questions. They compile a working list of questions "worth knowing" such as: When you read or observe something, how do you know what it means? Why do symbols change? What is progress? Leading with compelling questions as the driver for all learning embraces this backward design approach. While essential questions are mobilized in the UbD approach, the structure limits its capacities to build intellectual curiosities while maintaining responsiveness to the needs of learners. Through questions, students are given the tools to contemplate large-scale problems and develop ways to advocate for themselves and their communities. Teachers "behave oddly" by pressing students to interrogate their own and their classmates' assumptions and rarely offer information but rather divergent questions that shake the previous flow of classroom logic.

Using questions to teach science ethics

Questions drive critical, subversive, interdisciplinary, and self-aware discussions in classrooms and are therefore pivotal to disruptive STEM teaching and learning. In this section, I explore an example of how questions can foster conversations around medical ethics. Jillian Hahn is a public school biology teacher in a small rural district of Upstate NY. She developed a lesson that hinges on an essential question posed to learners: Should Louis Pasteur be considered a hero or villain? Jillian had found a book years previously in her basement, *Believing in Yourself*, published in 1976, that was part of a series by Johnson and Pileggi (1976) on values such as kindness, imagination, and determination. The book cheerfully and uncritically centers the protagonist Louis Pasteur, who perseveres despite public ridicule about finding an "invisible enemy." According to the story, after long hours in the lab, Dr. Pasteur was able to save a young child named Joseph Meister who was bitten by a rabid dog in rural France. The book paints a picture of Louis Pasteur saving a child by forming an army of soldiers who are sent into the body to fight off the evil rabies germs. The illustrations bring metaphor to life with a line of soldiers carrying rifles marching inside of a hypodermic needle. Engagement with this text shows students how Louis Pasteur has been heralded as one of the fathers of epidemiology for his groundbreaking work on vaccine development. Students are encouraged to examine how Louis Pasteur is portrayed in the piece and identify evidence that would support a claim that Louis Pasteur was either a hero or villain.

After discussion, the book is juxtaposed with a 13-minute video segment, entitled Rabid Roulette, from a TV series called *Dark Matters: Twisted but True* (Rosenthal & D'Innela, 2012). This series aired on the Science Channel in 2012–2013 and featured egregious ethical violations coupled with ominous music and a campy-true crime style. Episodes are still available via video streaming sites. The video clip reveals compelling evidence from science historians that Louis Pasteur had never experimented on human subjects before Meister and wasn't even qualified to administer the vaccination because of his background in chemistry rather than medicine. It details how Joseph Meister's last injection from Pasteur was a particularly potent dose of rabies to prove the vaccination provided immunity. In the clip, the classist nature of the interaction also comes to light. Joseph was from a poor, rural family, while Pasteur possessed some of the finest laboratory facilities in Paris. Interestingly, the biology students from Jillian's class immediately pointed out this distinction. One mentioned that Joey Meister travels by horse and carriage in the book while in the video clip he has to walk with his mother long distances to seek medical attention. By highlighting class differences, Jillian's rural students seem drawn into the storyline. After viewing the video clip, students are then asked once again if Pasteur should be considered a hero or a villain, using evidence from the video to support or counter their initial claim.

Understanding the histories behind the development of vaccines can offer insights into our current public reception of science. Ethical neglect by scientists of vulnerable populations such as children or poor and Black communities, to name only a few, is part of science that is often silenced in secondary school renditions. In the wake of the COVID-19 pandemic, ethical dilemmas of this sort continue to prevail. The new threat lies not in the hands of the scientists who are working diligently and collectively toward a resolution but in the misinformation that overwhelms our social media outlets. Public resistance toward vaccination has prevailed in recent times due to a series of ungrounded myths and misconceptions. The need for science literacy is now and teachers can lead this charge by providing an evocative curriculum that confronts the ethical implications of vaccines' discovery and distribution campaigns. Science informed citizenship and collective healing are inextricably connected as we move toward a post-pandemic, endemic era.

While clearly substantiated in the science community as necessary and effective, vaccination "controversies" continue to creep up in daily news feeds. In the rural communities in Upstate NY where I've worked and lived my entire life, vaccination rates are as low as 40% for COVID-19. During one of our conversations on the topic, Jillian Hahn added,

> It's good to talk to kids about vaccinations and how they work because I think there is still some fear about vaccinations even today. I've seen people posting on social media that they wouldn't get a coronavirus vaccine if one came out because it hasn't been tested. We've had vaccines for so long. What are you talking about? You'd rather risk dying or kill an elderly family member because you are afraid? They know how to make vaccines; this isn't something completely novel.

As a science educator, this signals the need for science teachers to initiate vigorous conversations and work harder to ensure that our students can make the best decisions for themselves as citizens.

It's important to walk the story of vaccination development back from present to its beginnings in the late 18th century in order to give a chronology of important discoveries in science. The history of vaccinations is a complex and multi-layered case study of how the science community approaches ethical dilemmas. For instance, smallpox was a devastating disease that was documented even during the era of Egyptian pharaohs. The first vaccination campaign was in response to smallpox outbreaks that wiped out a third of the people who were infected. This wave of infectious disease in the 18th century spread throughout the world, causing the most damage to the vulnerable, mainly young children. Scientist Edward Jenner discovered that people exposed to cowpox were immune to the smallpox infection (Science Friday, 2021, January 1).

Dr. Jenner made this revelation by purposefully infecting an 8-year-old child, James Phipps, first with cowpox and then, after recovery, to smallpox. Not only did the doctor expose this child to smallpox once, but he also repeated the process 20 times to ensure that his hypothesis was indeed correct (Science Friday, 2021, January 1). The child was his gardener's son and there was an obvious power differential at play. What choice did the gardener have when his financial stability was at risk? What if this experiment had gone badly and did not result in the striking positive results? Now, Jenner is revered as the father of vaccinations and supposedly worked the rest of his professional career as an advocate for vaccines. He continued to vaccinate children free of charge, especially children from impoverished backgrounds. The World Health Assembly deemed smallpox globally eradicated in 1980 (Center for Disease Control, 2017). Reconciling the fact that this single child's body was rendered disposable in order to instill so much good on the current and future world remains subject to ethical discussion. Shifts in thinking about bio-protections from ever-evolving pathogens are in fact the only constant (Markel, 2016).

I taught Hahn's lesson during remote instruction in Spring 2020 for a Theories of Teaching and Learning course as a means to address the current moment in the pandemic when the emergence of a COVID-19 vaccine was in its infancy. I had every student in my college-level course post a two-minute response on Flipgrid to the question of whether Pasteur was a hero or villain after reading the children's book and watching the *Dark Matters* clip. The Flipgrid platform allowed every student to have equal airtime as well as the ability to listen to their peers multiple times if needed and to respond thoughtfully. Student Flipgrid results skewed in the direction of deeming Pasteur a hero, which frankly surprised me. The primary line of reasoning my students drew from was that Pasteur was a risk taker and that the risk paid off. While my students didn't agree with the choices Pasteur made, they found them to be necessary in order to reach his goal of rabies vaccine discovery.

For me, this response demonstrated the type of scientific messaging that normalizes professional White, male dominance in science. Pasteur followed in the footsteps of Jenner, who has also been revered for his scientific contribution in a similar fashion. My class felt that Pasteur's actions lead to not just one isolated discovery but unlocked an entire scientific field of study, immunology, which therefore made his ethical transgressions worthwhile. Essentially, the consensus in the class was that unethical decisions that violate the humanity of a few are acceptable if the risks taken result in major scientific progress. As Jillian notes, "So there's a whole bio-ethics that you can talk about here. Now we have this vaccine, we saved our children's and our pet's lives but he could have KILLED this kid. How many other kids died?"

One undergraduate, Vinny, who tends to be rather reserved in physical classroom spaces, assumed a vocal role in this discussion through a strongly stated Flipgrid post that undoubtedly labeled Pasteur as a villain. Vinny said he did additional research to learn a bit more about Pasteur to further support his assertion.

Here is a snippet from Vinny's post:

Pasteur was actually a villain because he had no degree or medical license and he was doing tests on people without the actual correct procedures.

He lied to a lot of people on the ways he was doing his test such as he said he tested his vaccine on fifty dogs and cured them all from rabies but in reality it was only eleven dogs he actually tested out. This lack of information could have been harmful to people and caused more injury than good. He also sued the person who claimed to have discovered the first rabies shot in order to have for himself that title. It was never clear or pressing how he got the vaccines or medicines and so that's something that doctors do is try to inform people to help people's lives and not just to be the first one to put something new into the market. He was very much a villain.

The discussion then expanded to a whole group setting to hear the various student perspectives on whether or not Pasteur should be considered a hero or a villain. I did pose a follow-up question to the group, asking whether or not intent mattered. We thought about the intent of a discoverer and how it factored into the conversation. Pasteur received great notoriety for his efforts and this could have been an internal motivator for the risks taken as part of the rabies vaccine discovery process at the possible expense of Joseph Meister, who ironically later became a concierge at the *Institut Pasteur*.

Upon reflection, I would work to craft more intentional questions as part of the discussion and provide space for peer dialogue without direct teacher intervention. It was also of interest that students at both levels had never heard of Pasteur before despite being well aware of the term pasteurization. Some background could also be provided to set the students up to learn more about this historical scientific figure, especially at the high school level. I could see this lesson translating into a taking sides-type activity to visualize the contours of the argument across different gradients of agree, disagree, or undecided.

As an outcome of the lesson, students were able to visualize more clearly the human side of scientific discovery and the grayness of ethical issues in science. Students contemplated their own moral positions in relationship to how they engage with the world and the decisions they make in their own lives. There are a myriad of other examples that involve the unethical practice of science on the bodies of Black and Indigenous people and other people of color. There is much work that could potentially be done in this arena to counter a predominant narrative of science as infallible.

The critical analysis of texts makes the Pasteur lesson particularly disruptive. The lesson walks learners through the process of identifying language choices that signify bias. It encourages them to think beyond one perspective and a single narrative view. The lesson also subverts traditional STEM instruction by featuring non-traditional texts such as a picture book and TV programming. The science content surrounding vaccines is revealed through an interdisciplinary focus on ethics as well as history. Vaccination is a topic that clearly involves the situated self, as we all make choices about our own health that can impact those around us. The availability of COVID-19 or Flu vaccines around the world can lead to a conversation with students about the global distribution of resources and political leverage.

Societally, We've moved well beyond injecting small children with a live virus without consent. We have processes in place to protect test subjects and ways to validate the efficacy of a vaccination in precise and replicable terms. Now that vaccines are available for wide-scale use, a marker of wealth and global status, the US vaccination rate remains the lowest out of the G7 nations (Lukpat, 2021).

Anti-vaccination advocates offer a series of appealing narratives in an effort to sway public opinion, including safety issues, conspiracy theories, and alternative medicines, as well as about the origins and remedies to COVID-19, often leveraging historical ethical atrocities. In their investigation of competing views of vaccinations as portrayed on Facebook, Johnson et al. (2020) found that "although smaller in overall size, anti-vaccination clusters manage to become highly entangled with undecided clusters in the main online network, whereas pro-vaccination clusters are more peripheral" (p. 230). The social media presence of anti-vaxxers is aggressive and unyielding. Roose (2020) from the *New York Times* mentions how the responsive and streamlined development of the vaccination offers fuel to a movement looking to poke holes in the credibility of the science and the final product. The organization and persistence of the anti-vaccinator movement does not bode well for future booster vaccination campaigns. While we want to foster a healthy skepticism of scientific discovery in our curriculum, we need to draw from credible sources and a robust knowledge base in order forge critical media and science literacies.

Planning for digital futures

In order to critically view disinformation on social media, secondary-level teachers, in particular, need to equip learners with the tools to spot manufactured narratives and fake news. I argue that critical media literacy skills are part of future STEM literacies and therefore must be thoughtfully considered as part of the curriculum creation process. A barrage of questions are associated with technology use that shape both the ways we teach and learn. Technology is expansive as a term, but for the purposes of this chapter, will discuss implications of its use as a tool at both the classroom and school system level. AI is no longer on the horizon but here in our backyard. Dialogue regarding the ethical use of AI is a matter of urgency.

Within the past three to five generations alone, massive societal changes in the use and consumption of digital technologies have occurred. The COVID-19 pandemic only reinforced this reliance on technology in schools. Teachers face shifting digital landscapes and an urgency to lean on tech. Postman and Weingartner (1969) also caution that the world that students encounter as adults is vastly different than the world of today. They make an urgent call for educators to steer away from recall and memorization and instead to offer thinking opportunities for navigating the future. While the authors were referring to the technological advances of the time, such as use of TV/media for teaching, this issue of anticipation of future worlds endures.

Jeremy, a seasoned eighth-grade science teacher at a suburban school district in the Northeast, discussed this movement away from fact-based practices. Fifteen years ago, Jeremy admitted that part of his teaching bravado reinforced the skill of memorization. For instance, he would "wow" the class by memorizing the distance of Earth from the planet Mars. Now, students are able to access this information readily through the use of 1:1 Chromebooks provided by the district as well as personal devices.

[Technology is] another tool in the arsenal to enable you to teach. That has recently shifted for me. I used to really ascribe to that belief system... I think a fundamental shift happened once the kids went one to one with online devices where I was not the source of information anymore and the information could be delivered to the student in more efficient ways than I was able to deliver it...I can't wow them with facts anymore. But I can set the stage for cool inferences.

(Jeremy, interview 4/12/16)

Jeremy recalls this moment as palpable when his pedagogical outlook shifted more in favor of the learner than the performativity of the teacher. Jeremy recognized that this intellectual shift was quite positive for his students, but there seemed a tone of nostalgia for the past due to its ease and familiarity.

Jeremy applied technology not only to boost student interest but also to increase the potential for greater accessibility through both visual and auditory modalities. Computer access allows the learner to view websites and interactive simulations that incorporate both video and audio information. For instance, Jeremy's class referenced an online periodic table that offered 3-D views of atomic structures and audio pronunciation of element names. Jeremy used to have students create posters only as an alternative assessment. His classes currently use various software packages to make student projects, now the featured assessment, come to life with pictures and videos. He recalled, "Now that the digital media has come around it's kind of become something more interactive" (Observation, 4/12/16). As part of science teaching, digital technologies are integrated as part of almost every lesson and include the use of social media to post outcomes more publicly.

The emphasis on digital learning comes with social costs, which educators will have to sort out to fully understand its implications for classroom dynamics. In Jeremy's district, there is open access to Wi-Fi, and students have essentially all day access to the internet, which at times they use recreationally. There are mixed messages socially about how to engage within the classroom community. Jeremy and other eighth-grade teachers went to the extent of acting out various social scenarios to model appropriate behaviors. They spend time working with the eighth-grade students on public speaking and encourage them to share lab outcomes within small and large group settings of around 100 students in total. Jeremy and his teacher team also encourage behavioral norms such as facing each other while speaking. Yet, the consistent use of the Chromebooks dramatically interferes with social etiquette because students are tethered to a screen. When students enter the classroom, they have morning work on their Chromebooks and almost exclusively use them to upload work. One of the teaching assistants for the classroom created an end of the year slideshow. The slide show presented 17 images, each projected for a total of three seconds before moving on to the next. In the background was contemporary music, like the kind you would hear on a latest hits radio station; 2 of the 17 images depicted students looking at computer screens. This comprised 10% of the slideshow and seemed symbolic of the normalization of digital technology as part of the social experience of school.

With the rollout of the Common Core, tech companies poised to deliver standards-based curriculum flourished. Now, after years of pandemic era teaching, big tech as an industrial complex stands to gain at an astounding rate. Kardaras (2016) brings to light the extent of corporatization that has occurred in order for students to have access to now common classroom devices such as tablets. He refers to once egregious example from New York City where school chancellor Joel Klein concocted a plan to roll out an Achievement Reporting and Innovation System (ARIS) that supposedly tracked students and gathered data. When put into motion, the ARIS failed to function at the level of expectation and cost NYC schools millions of dollars. Klein then left public service to assume a role at the tech company that he himself had hired to maintain the ARIS system. Unfortunately, there are many other examples that could be included of money squandered through the introduction of devices, software, and annual licenses in schools that desperately need funds for basic student needs.

Jeremy's dynamic experiences with technology over the course of his career indicate that he embraces it as a tool that can ignite further learning, "I can set the stage for cool inferences." This mindset aligns with large-scale analysis of teaching with technology that found that the most effective inclusion is facilitated by a high-quality, committed teacher. Technology is a vessel to elevate student motivation that enhances instruction. Like any tool of education, it needs to be incorporated with care in order to be sustained (Higgins, Xiao, & Katsipataki, 2012).

Another emerging technology tool in education that sparks much debate is the use of generative AI. The implications of AI to the field of education remain contested. The December 9, 2022, Atlantic article by Daniel Herman entitled *The End of High School English* describes the potentially devastating effects of the AI program ChatGPT. Based on his initial use of ChatGPT, he found that it wrote in a sophisticated enough way to answer secondary-level prompts with proficiency. Herman, who taught English language arts (ELA) for a 12-year-period, observed a tracking system in place, much like in mathematics. Lowest in the hierarchy included students exposed primarily to basic comprehension, grammar, and writing structures. In the middle tier, students practiced essay construction and argumentation. The highest-tier students spent classtime analyzing tone, genre, and rhythm of writing. The introduction of ChatGPT gave Herman cautionary pause and left him to wonder how this technology may shift content and pedagogical focus.

Greene (2022), also with years of ELA teaching experience, takes a more positive outlook, but admits that many teachers "will have to do some soul searching." Greene conveys excitement for this new juncture that allows for secondary-level writing instruction to be reimagined in divergence from prior structures that rewarded mechanical responses in order to perform satisfactorily on standardized tests. "ChatGPT doesn't mark the end of high school English class, but it can mark the end of formulaic, mediocre writing performance as a goal for students and teachers." Postman and Weingartner (1969) also break down subject area teaching into two camps, "old" and "new." New teachers desire to teach content and rely on conceptual understanding to solidify learning. Using English teaching as an example, the "old" focus would be on grammar and the "new" broadening the notion of text to include nearly any social interaction and using the tools of inquiry to develop complex understanding of language and humankind.

AI at a systems level

Schools are actively complicit in the use of technology that relies on racialized and biased perceptions. With the climate of fear associated with the recent onslaught of school shootings in the United States, districts turn to technology as a means of false security. Schools actually function as accessories to law enforcement with common computer networks to surveil communities of color. As an example, the school to prison pipeline is so intimately connected that last year that the police department of Syracuse, New York, in the Upstate region of the state reported that a portion of 430 crime cameras that line the streets went down when the Syracuse City School District's computer system was hacked. This outage resulted in gaps of surveillance footage associated with two major incidents in the city. Mary Kielar of CNY Central reported that the district had to shell out substantial funds to get access back, but the numbers were not specifically known. The implications of 24-h surveillance in disinvested communities are strongly reminiscent of Foucault's explanation of the Panopticon as a tool of societal control. Panopticon is a prison structure where the individual cells are easily observed from the security tower, while the corrections officers are not visible but assumed to be omnipresent. Foucault (1977) extends this concept as a larger metaphor for societal function where control of citizens is largely maintained through internalized authority structures. Through this lens, omniscient surveillance technology acts as the corrections officer poised to quell any uprising by rowdy citizens, including our children placed in the care of public schools under the guise of education.

Scholars such as Dr. Ruha Benjamin provide insights on how technology also fuels White supremacist agendas and perpetuates inequity. Benjamin's (2019) book, *Race after Technology: Abolitionist Tools for the New Jim Code*, coins the term as the latest way Black and Brown bodies are oppressed by society. "The employment of new technologies that reflect and reproduce existing inequities but that are promoted and perceived as more objective or progressive than the discriminatory systems of a previous era" (p. 6). Building on Michelle Alexander's (2010) pivotal work, *The New Jim Crow*, Dr. Benjamin argues that the technologies we rely on to make daily life easier, identify instances of criminality, and stay connected are actually putting into use racist algorithms that devalue BIPOC lives and actively target them for incarceration.

The Algorithmic Justice League (AJL), founded by Joy Buolamwini, aims to "empower communities and galvanize decision makers to take action that mitigates the harms and biases of AI" (https://www.ajl.org/about). AJL advocates for equitable and accountable AI as a broader cultural movement for systems-level change. Equitable AI requires that people have control or affirmative consent over the use of their bodily information so that it does not become a tool of dominance or suppression. Equitable AI also stresses that people have knowledge of how AI is being used in their daily encounters and the potential risks and harms associated with its use. Furthermore, to preserve the humanity of communities regardless of racial identity, AI should be removed from the equation in cases of lethal force or in other instances in the public sector where racial profiling is a rampant. Accountable AI as outlined by the AJL focuses on three primary factors.

Meaningful transparency involves active communication on the use of AI and its purpose. Facial recognition software companies and consumers need code of conduct protocols in place as visible commitments to respect all lives. Continuous oversight by companies and governments that opt to use AI software is necessary to regulate the interactions between technology and people at the greatest level. A process of documentation and auditing is needed as one layer of protection. Lastly, AJL requests that harms against people of color as a result of the use of AI be acknowledged, and some form of restorative action taken. Through the critical facet of disruptive STEM teaching, media literacy skills can be imparted on secondary-level learners, enabling them to gain awareness of the stakes involved with the large-scale use of AI and ways to advocate for its equitable use.

Implications of technology and planning for disruptive stem

Learners have opinions and insights to foreground in classroom settings regardless of the preconceived plan. Posing essential questions equips secondary students to engage in future world-making with an informed and proactive mentality. Ethical dilemmas in STEM abound and can anchor curriculum by providing space to work through complexities in a scaffolded manner. Tracing historical lines of social inequity based on race and class difference can shed light on social phenomena of today. I argue that we need not shelter our adolescents, waiting for another day to broach difficult subjects, but rather trust in their capacities to think through complex global issues of present. If we fail to include such topics as AI in our curricular considerations in STEM education, we miss out on a level of preparation that students ultimately deserve as current and future tech consumers. Technology supports students in differentiated ways to satisfy their academic goals, including students with disability labels. Approaching mathematical skill-building through the angle of computer science invites students to strengthen their number sense and logic, since the representation of numbers and algorithmic applications are foundational to the field of computer science. AI presses us to consider questions worth knowing, such as "what makes us human?" and "what world do we want to live in?"

References

Alexander, M. (2010). *The new Jim Crow: Mass incarceration in the age of colorblindness.* The New Press.

Benjamin, R. (2019). Race after technology: Abolitionist tools for the new Jim code. *Social Forces* https://doi.org/10.1093/sf/soz162

Center for Disease Control. (2017). *Small pox.* https://www.cdc.gov/smallpox/index.html.

Delpit, L. D. (2012). *"Multiplication is for white people": Raising expectations for other people's children*. The New Press.

Foucault, Mi (1977). *Discipline and punish: The birth of the prison*. Pantheon Books. https://monoskop.org/images/4/43/Foucault_Michel_Discipline_and_Punish_The_Birth_of_the_Prison_1977_1995.pdf

Greene, P. (December 11, 2022). No, ChatGPT is not the end of high school English. But here's the useful tool it offers teachers. *Forbes*. https://www.forbes.com/sites/petergreene/2022/12/11/no-chatgpt-is-not-the-end-of-high-school-english-but-heres-the-useful-tool-it-offers-teachers/?sh=117981f41437

Higgins, S., Xiao, Z., & Katsipataki, M. (2012). *The impact of digital technology on learning: A summary for the education endowment foundation. Full report*. Education Endowment Foundation. https://larrycuban.wordpress.com/wp-content/uploads/2013/12/the_impact_of_digital_technologies_on_learning_full_report_2012.pdf

Johnson, N. F., Velásquez, N., Restrepo, N. J., Leahy, R., Gabriel, N., El Oud, S., Zheng M, Manrique P, Wuchty S, & Lupu, Y. (2020). The online competition between pro-and anti-vaccination views. *Nature, 582*(7811), 230–233. https://doi.org/10.1038/s41586-020-2281-1

Johnson, S., & Pileggi, S. (1976) *The power of believing in yourself*. Value Communications.

Kardaras, N. (2016). *Glow kids: How screen addiction is hijacking our kids-and how to break the trance*. St. Martin's Press.

Lukpat, A. (2021, September 21). The US is falling to the lowest vaccination rate of G7 nations. *New York Times*. https://www.nytimes.com/2021/09/11/world/asia/us-vaccination-rate-low.html

Markel, H. (2016, July 7). Louis Pasteur's risky move to save a boy from almost certain death. *PBS New Hour*. https://www.pbs.org/newshour/health/louis-pasteurs-risky-move-to-save-a-boy-from-almost-certain-death

McCreary, M. (2022). Beyond backward design, or, by the end of this article, you should be able to imagine some alternatives to learning objectives. *To Improve the Academy: A Journal of Educational Development, 41*(1). https://doi.org/10.3998/tia.454

McTighe, J., & Wiggins, G. (2013). *Essential questions: Opening doors to student understanding*. ASCD.

Pendleton-Jullian, A. M., & Brown, J. S. (2018). *Design Unbound: Designing for Emergence in a White Water World, Volume 1: Designing for Emergence* (Vol. 1). MIT Press.

Postman, N., & Weingartner C. (1969). *Teaching as a subversive activity*. Delta.

Roose, A. (2020, May 13). Get ready for a Covid-19 vaccine information war. *New York Times*. https://www.nytimes.com/2020/05/13/technology/coronavirus-vaccine-disinformation.html

Rosenthal, A. & D'Innela, A. 2012, July 21. *Rabid Roulette (2, 2) In executive producers Dan Gold, Jasper James, Dark Matters: Twisted but True*. Science Channel.

Science Friday. (2021, January 1). *Where did the word vaccine come from?* jhttps://www.sciencefriday.com/segments/science-diction-vaccine-history/.

Wiggins, G., Wiggins, G. P., & McTighe, J. (2005). *Understanding by design*. ASCD. https://doi.org/10.14483/calj.v19n1.11490

Content-focused, disruptive STEM curriculum, with specific in-depth lessons

3
UNDERSTANDING ECO-COLONIALISM THROUGH GLOBALLY COMPETENT STEM PRAXIS

Margery Gardner

Introduction

There are two unfolding storylines that comingle in this chapter. The first storyline describes the experiences of David, a Colombian American student teacher, as he develops critical practices and greater reflective capacities. The second describes my own challenge to bring into focus the concept of eco-colonialism through a tangible lesson. Both storylines are conceptually rooted in the notion of globally competent disruptive STEM teaching. The disruptive STEM facet of interdisciplinarity is most prominent, since it allows for the combination of science-based content on agriculture alongside political and economic considerations.

As globalization actively plays out in our schools, curriculum and instructional considerations should respond accordingly to locate opportunities to blend and uncover historical contexts, international politics, and global environmental concerns. Competencies to navigate globalization through multidimensional sociocultural and scientific lenses require explicit and thoughtful engagements in school that fall within and beyond the purview of traditional disciplinary silos. Tichnor-Wagner, Parkhouse, Glazier, and Cain (2019) describe: "Global competence is the set of knowledge, skills, mindsets, and values needed to thrive in a diverse, globalized society" (p. 3). Terms such as *civic competence, global awareness, global citizenship, global literacy, intercultural competence, international education,* and *global education* have also been used in prior literature as referents for this concept.

Disruptive STEM education situates the self as part of a more expansive experience that involves understanding ones' positionality in the world.

DOI: 10.4324/9781003395782-6

Tichnor-Wagner et al. (2019) explain globally competent teaching as a means to foster and facilitate knowledge, dispositions, and skills that enable students to fully engage with and within globalized contexts. Knowledge of complex global issues arises through opportunities for analysis and critique. Dispositional competencies such as empathy, global outlooks that embrace multiple perspectives, and introspective investigations of identity and positionality are other vital components. Skills such as cross-cultural communication and respect for language, specifically mother tongues, and advocacy for justice are also included in this vision.

Teacher candidates or pre-service teachers face heightened preparation demands. A prioritization process occurs at both the programmatic level and through individual course selection that provides insights on the valuation of global citizenship. Poole and Russell III (2015) surveyed over 200 teacher candidates at a large university in a southeastern state. The research team found that 40% of respondents had never taken a course on global/international issues as part of their preparation, while another 26% reported only one course. Around 43% of teacher candidates reported that they had never attended a global/international lecture during their entire 4-year program. About one-third either watched or read global news, but not on a consistent basis.

Poole and Russell III's study (2015) verified the need for global content representation in teacher preparation coursework since several previous studies found a high correlation between teacher candidates' exposure and the enactment of globally competent curriculum in the P-12 sector. If teacher candidates are not provided with global perspectives during their preparation, then the likelihood of their incorporating global themes within P-12 classrooms sharply diminishes. Some of the characteristics described as crucial to global competence include awareness of the state of the planet, awareness of human choices, and knowledge of global dynamics (Hanvey, 1976; Poole & Russell III, 2015). Cross-cultural experiences, such as study abroad opportunities, service learning, or other university sponsored events support the development of teacher candidates' global perspectives and empathy for others.

Environmental issues are among the most important challenges that globally competent STEM teaching must address. Global citizenship necessitates an understanding of the critical zone, or the thin layer of the Earth where humans primarily exist that consists of vegetation, soils, waterways, and other animal life. In order for communities to make environmentally informed choices about local water and soil resources, ecosystem services, and community climate resilience, they must be able to make sense of data and be able to manipulate local and national data sources to interpret impacts of human activity. There are major gaps in coverage of the critical zone across K-16-Education. Currently, uneven access to critical zone educational opportunities stymies progress toward understanding "grand challenges" (White et al., 2017) like climate change. Scherrer (2022) surfaces the damage of teaching climate science in isolation from emotional, historical, and cultural attachments to spaces that are fully racialized. Disruptive STEM education takes into consideration the need to bring dimensionality in order to understand complex global issues and catalyze interest and concern from secondary-level learners.

In this chapter, I present ways to actualize globally competent teaching through discussions of eco-colonization in Central and South America. A critical lens illuminates the violence inflicted on Indigenous communities and their environments through corporate land grabs and political corruption. I highlight the experience of Colombian American teacher candidate David, as he disrupts negative political tropes about immigration in his classroom. David majored in history throughout his undergraduate years and is therefore particularly equipped to demonstrate the pragmatics of globally competent teaching. Through his own personhood as a Colombian American, his situatedness undergirded many of his pedagogical decisions that bring to the surface Latin American political struggle. This chapter also explores the possibilities of teaching about the harms of monocultural farming in a secondary science setting through curricular exemplars.

Bringing Latin American issues into focus: One student teacher's experience

I worked with a student teacher named David who sought a certification in middle and high school history/social studies. David had a bit of a rocky start in the first weeks of his placement. He felt caught between the world of theory and actualization. The school district was close to the university that David attended and, therefore, was riddled with class divides between local rural students and faculty members' children. The classroom where David student-taught was cramped. The 11th-grade section held over 30 full-grown bodies, so opportunities for group work were limited by the sheer spatial parameters of the room. To further exacerbate this situation, David's own educational experiences at the secondary level were far from innovative. David even described how at one point his teachers just put on a movie and had the TV teach the class a series of lessons. Yet, David's eagerness to connect to his students and learn about them as whole people gleamed through this less than perfect teaching situation.

Around halfway through the semester, David had a bit of an epiphany when it came to his teaching. For the first several lessons that I observed,

David mainly lectured while students took notes. While an effective practice in moderation, David realized that there was much room to grow. Reflecting on the experience, he said:

> At first I didn't get the essential question thing until half way through student teaching. I then asked essential questions that I found were very important [in order to] break up [the lesson] into different parts. Based on these questions, here are the different topics. Early on my lesson plans were 1/3 of a page but by the end they were 2 pages. If the conversation stalled I would have additional resources.
>
> *(Interview, 12/16/20)*

This shift, that may seem outwardly modest, offers a glimpse into the complex journey of becoming a disruptive STEM teacher.

Once David was able to anchor his lesson on a thought-provoking topic, the shift in student response was nearly immediate. This shift in pedagogical approach closely connects to the discussion of "questions worth knowing" in Chapter 3 of this book. Postman and Weingartner (1969) assert that question posing is a crucial component of subversive teaching that offers students opportunities for critical thought and empowerment. In subsequent observations, I noticed how much more visibly engaged the class seemed in comparison to previous observations. David also showed vulnerability by eliciting feedback from the class about his question-posing pedagogy. During this relatively short period of time, David learned to prioritize the voices and needs of his students above demands from the outside.

David mentioned in our reflective interview that Latin American politics aren't typically taught because there are so few questions about the region on the end of the year standardized test. In many ways his narrative reflects the facet of subversion by pulling to the forefront a storyline often left behind or covered only in cursory detail. David identifies as "half Colombian" since his mother immigrated from South America. He felt a strong affinity to Colombia and spent a lot of time there with his family while growing up. He recalls:

> One thing that is really clear when you are spending time in Colombia, if you ever worked for the US government you cannot say it because you will be shut out of every conversation. We had a terrible history in Colombia and honestly every Latin American country. To be honest, the US has pretty much had a position of constantly seeing the Western Hemisphere as its playground where it can do what it wants. [The US] repeatedly overthrows governments to get their way.
>
> *(Interview, 12/16/20)*

David channeled his heritage as a Colombian American and a familiar critical perspective on US intervention to reframe the issue of immigration for the learners in his class. He conveyed through his lesson how historical interventions set the context for displacement that we see today.

According to Loewen's (2008) *Lies My Teacher Told Me*, Woodrow Wilson authorized intensive interventions in Latin American during his presidency. "We landed troops in Mexico in 1914, Haiti in 1915, the Dominican Republic in 1916, Mexico again in 1916 (and nine more times before the end of Wilson's presidency), Cuba in 1917, and Panama in 1918" (p. 16). David constructed a lesson that counterbalanced the dominant narrative in the United States that absolves the government of any wrongdoing.

David opened the lesson with a philosophical question for the class: "If someone has done harm then is it possible to undo the harm that you do?" After some discussion, he asked the class to think about how governments can undo past harms. He probed further and asked students to think about legacies of slavery as well as racial, class, and gender privileges. David transitioned the discussion toward South American political exchanges, explaining that 71% of undocumented migrants to the United States come from Mexico and Guatemala. He followed up with, "Why do they have so much violence and unrest in these countries?" A student responded, "Us." After providing background knowledge to the class, David mentioned that he is proud to be an American, but he wants to highlight some of the bad things that our country has done. "We need to seek to undo some of these wrongs because they perpetuate today." As a class they read aloud Chilean poet and politician Pablo Neruda's *La United Fruit Co.* from Canto General (2000), which sheds light on the exploitation of Central American countries for the benefit of the United States. David then posed another question about current immigration policies: "Why don't people come in legally?" He followed up with an explanation of the extreme difficulties involved with immigrating to the United States through legal pathways. David shared the stunning statistics that 26,000 people from Mexico are let in each year and 1.3 million have been on the waitlist since 1997. David wrapped the lesson up by initiating discussion once again on the central question regarding harm.

David's evolving practice offers insights on how globally competent teaching's core competencies of knowledge, dispositions, and skills can be embedded. David centered Latin American politics in this lesson on immigration that pulled from existing and new knowledge of patterns of migration juxtaposed with political strife. He asked students to challenge their own ideas about Latin American immigration through rhetorical analysis of different primary sources. By pulling the lesson together using the emotional thread of harm, he is able to evoke a dispositional competency of empathy. Since the learners in the class were reaching voting age,

David encouraged his class to register to vote and make informed decisions when selecting representatives.

Banana republics and the making of monoculture

STEM education is not immune to political context but rather provides a backdrop to help learners understand how knowledge is generated and shared. Teaching through a disruptive lens identifies paths that make clear these sociocultural milieus. Science knowledge is frequently exploited for capitalist gain, resulting in harm to our planet. Disruptive STEM can embrace a globally competent approach to give greater meaning to our curriculum that promotes advocacy for environmental and social justice. David's lesson ignited my own curiosity about how eco-colonialism could be further explored at the secondary level. In the subsequent section, I provide background information on eco-colonialism as it pertains to banana monocultures in Latin America and share my own rendition of a secondary-level lesson.

Banana republic is a condescending term devised by the United States for countries that rely heavily on commercial export and that afford corporations partiality at the cost of their own autonomy, particularly those in Central and South America. These countries are deemed politically volatile by the United States and are seen as reliant on large agribusiness in order to function. Beginning in 1899, the United Fruit Company, whose headquarters were located in Boston, MA, started operations in what is now Belize (Moberg, 1996). The United Fruit Company rapidly became the largest exporter of bananas to North America and actively sought to eradicate any competitors who threatened their large profit margins. By the 1930s, the company's land holdings in Central America were immense. Through a combination of bribery and extortion, Central American governments allowed the exploitation of their lands by United Fruit Company. In 1954, United Fruit Company conspired with the Central Intelligence Agency (CIA) in the United States to stage a coup in Guatemala. This coup catalyzed years of political unrest in the area that eventually cascaded to other countries like El Salvador. Civil War in the area caused the genocide of Indigenous people – 75,000 in El Salvador and 200,000 in Guatemala. As Jessica Hernandez writes, "the United States created political turmoil in our countries because they wanted to support the political leaders who were in favor of further privatizing the lands in our countries to sell to these foreign corporations" (2022, p. 192). In its current iteration, the United Fruit Company is now Chiquita, and in 2014 it merged with Irish company Fyffes. Alongside Dole and Fresh Del Monte, these companies control 80% of the banana production market (Berman, 2014). According to the

United Nations, bananas remain the largest fruit trade in the world, with a value of about 8 billion USD in 2016.

Indigenous scientist Jessica Hernandez (2022) conceptualizes this harm as eco-colonialism, which renders visible the impact of settler colonialism on natural landscapes. Eco-colonialism poses a direct threat to both the environment and native cultures, a historical force that persists today. Settler colonialism, which undergirds eco-colonialism, is an ongoing project of oppression where power to influence political systems is bestowed upon the settler. Settlers flex this power through their decisions about Indigenous people and their lands. This system creates capitalist opportunities that run in direct opposition to Indigenous cultures and knowledge systems. Care of land becomes usurped by desires for profit. As a result, Indigenous people, along with biodiversity of land and sea, become relegated to the margins.

The banana industry in Latin America demonstrates the impact of eco-colonialism through monocultural farming. Until the 1950s, the Gros Michel or "Big Mike" banana variety was highly popular across the globe for its sweet flavor and thin skin for durable shipping. The primary method of fruit propagation for the Gros Michel commercial banana crops was clonal suckers. Only cuttings from the plants with optimal yield and attractive attributes were planted. As a result, a monoculture was created with genetically identical fruits. As early as 1890, the fungal pathogen *Fusarium oxysporum f.sp. cubense*, or Panama disease, began to wipe out banana plantations that contained genetically identical Gros Michels. The fungus lingered in the soil for upwards of a decade, leading to the desertion of once-thriving banana-producing areas. As a result of massive single-crop plantations, outbreaks of Panama disease occurred throughout the 20th century. The flavor and appearance of another variety, the Cavendish, lacked in comparison, but with a massive media campaign, the United Fruit Company was able to establish a flourishing market for the new banana in the 1950s to replace the Gros Michel. "Big Mike" is now nearly extinct.

The Cavendish is now the banana readily available in grocery stores throughout the United States, representing nearly 99% of the export market. Like the Gros Michel, it remains a genetic clone of its former siblings. As a result, the threat of massive crop failure continues today, referred to by Paul Tullis of the Washington Post as "Bananapocalypse." There is a strain of fungus, officially termed "Fusarium Wilt Tropical Race 4 (TR4)," that has hampered banana farming in Asia since the 1990s. There is now a new strain of TR4 reported that attacks the roots and vascular system of the banana plant, including commercial varieties that are distributed in mass quantities across the globe. This strain, discovered in Malaysia in 1992/1993, appears to be migrating to South East Asia and beyond. In 2013, TR4 was found in Mozambique, followed by Lebanon and Pakistan by 2015, impacting an estimated 2.5 million plants. In response, countries such as Australia have banned banana imports to hopefully quell the spread. These embargos don't just impact large companies, but they also threaten the livelihoods of small-scale farmers in these regions.

Despite containment attempts, the fungus looks to be continuing to spread. According to *Washington Post* reporter Roberto Ferdman, the banana extinction process takes upwards of 50 years. During that span, there are hopes that a successful alternative or a disease-resistant strain can take root. But significant moral and scientific questions arise when considering genetically modified banana varieties. At present, TR4 seems to only be attacking commercial bananas and not 85% of the varieties locally eaten. If TR4 is able to impact other local varieties, then food security is at stake for regions such as East and Southern Africa. This case study provides evidence of the global impact corporate decisions have on the stability of our food sources. There exists a global imperative to understand how our food systems function so they can serve more equitably.

Disruptive STEM teaching as globally competent

The goals of globally competent teaching and disruptive STEM coalesce to illustrate how sociopolitical contexts bring about change to natural systems. Bybee (2010) suggested anchoring STEM-integrated curriculum around major societal challenges such as energy efficiency, resource use, environmental quality, and hazard mitigation that provide natural nodes of connection between disciplines. Through authentic problem/project-based learning, students can develop capacities to confront global issues and develop practical long-term solutions. Monocultures act as a topic that can be understood from multiple angles that has potential for redesign.

I argue that STEM classrooms should be sites for critical education and that learners are more engaged through sociopolitical contextualization. In order to bring structure and clarity to the lesson, I mirrored materials using Bybee's 5E planning model that is a sequencing approach now commonly adopted in science education. The 5E model begins with an entry into the concept of study through a high-interest hook called the "engage" phase. Bybee's model then breaks with teaching tradition through an "explore" phase that allows for open, firsthand exploration of the concept prior to any formal explanation. This phase gives all students in the class a common set of experiences that can act as fodder for future discussions, alleviating equity issues that can come about based on prior backgrounds. The next phase, "explain," gives students additional information about the concept at hand that cannot be rendered visible through experimentation alone. This information can color their future interactions with the material and add eloquence to their responses. The "elaboration" phase gives students an opportunity to extend the knowledge acquired into further realms or applications. Students can identify ways that the concept connects and transfers into different contexts, which allows students to splinter the concepts into areas of their individual interests. "Evaluation" is the final phase and allows the teacher to gather evidence of learning from their students. Evaluation should be open to different interpretations and assume different product forms. The phases of the model can be divided as necessary and oriented in a circular fashion rather than in a linear sort of way. Students are free to work through the phases at their own pace and review points of confusion by returning to a prior "E" (Bybee et al., 2006).

This model allows learners greater agency to understand monoculture farming and eco-colonialism in sophisticated ways and with contextualization. The learning goals support the understanding that monoculture farming practices make crops more susceptible to large-scale damage from disease and that monocultures are a global concern from both environmental and cultural standpoints. The "question worth knowing" extends from David's initial question: How do monocultures cause harm to both people and the environment?

A *Engage*

Learners listen to a song created in 1923: Yes! We Have No Bananas with the lyrics accessible in written form in both English and Spanish. Instead of telling students the background of the song, a conversation about the central premise represented in the lyrics can be generated, as well as predictions around the shortage of bananas.

B *Explore*

Learners then explore the origins and diversity of bananas as a food source from Africa. There are online resources and video clips available that describe the wide array of banana shapes, sizes, and flavors. If possible, bring in a sample of bananas to compare. Learners will write down their observations and share findings across groups.

C *Explain*

This stage would include developing a working definition for the term "eco-colonialism" using language responsive to the learners. In addition, the teacher will present reasons for Big Mike going "industrially extinct" (explain and emphasize the concept of "monoculture farming"), the current threat to Cavendish bananas, which many of the students may have just dissected, and why Panama disease is a global concern. Learners consider the reasons for the adoption of monoculture farming practice from a capitalist lens. The teacher should thus be prepared to hold space for explanations of social, economic, and political reasons behind monoculture farming worldwide.

D *Elaborate*

Learners investigate the latest fungal outbreaks using resources from the Centre for Agriculture and Bioscience International (CABI, sometimes also referred to as CAB International), such as an interactive world map showing the spread of TR4 worldwide (https://www.cabi.org/isc/tr4). Learners can then compare territories of Indigenous people using the online mapping resource Native Land Digital http://Native-Land.ca to assist in understanding colonial boundaries and land.

They will be asked to use the new knowledge learned in the "explain" stage to share trends and new understandings through these visualizations.

E *Evaluate*

Finally, students will work in groups to plan ways to challenge the existing monocultural system and replace it with new designs that embrace biodiversity and decolonial thinking. Learners are encouraged to make connections to their own cultural experience when designing. They should also consider the social, economic, and political advantages and concerns that may potentially exist if farmers adopt their model. Learners should present their designs to the whole class for feedback.

There is an opportunity within this lesson to view science knowledge as inextricably linked to indigeneity. A goal of the lesson is for learners to think "glocally" about their own local surroundings and intersecting identities in relation to global issues and perspectives to which we remain interconnected.

Assessment models that can provide adequate feedback on learning performances can be tricky, given the complexities of globally competent teaching. Tichnor-Wagner et al. (2019) suggest the use of multiple assessment measures that are authentic in nature to understand learner progress toward global competence. Tichnor-Wagner et al. (2019) locate assessment as part of a learning partnership with the purpose of driving instruction in meaningful ways. Folding criticality into the iterative learning and assessment process is a notable consideration. Without a critical exploratory lens, students may gain a superficial awareness of global issues that neglects systems and their influence on society and environment.

Learners deserve to confront complex global issues, including the impacts of eco-colonialism on Indigenous people and natural spaces. Such lessons are emblematic of disruptive STEM's core aspects that strive to build criticality, interdisciplinarity, and involvement of situated self. The legacies of harm associated with political and economic exploitation in Latin America for more than a century are still visible today as an extreme example of environmental and humanitarian disregard. STEM disruptors may want to collaborate with a social studies teacher like David in order to best subvert curriculum to embrace a historical perspective that provides insights on corporate transgressions today. There are wonderful resources available, including primary sources from the United Fruit Company advertising and tech tools such as a TR4 fungus tracker, which can further assist engagement with this issue. David demonstrated vulnerability as a student teacher from a marginalized Colombian American identity to bring to light the topics of immigration and corporate harms for his secondary students. Scherrer (2022) advocates for storied pedagogies of place that intertwine Black and Indigenous narrations of environmental care and kinship. As educators, we can instill new logics that support a future that prioritizes Indigenous healing and collective care.

References

Berman, G. (2014, March 10). *Next chapter in the global banana trade's bloody history: Walmartization.* Huffington Post. http://www.huffingtonpost.com/2014/03/10/worlds-largest-banana-company-_n_4935955.html

Bybee, R. W. (2010). Advancing STEM education: A 2020 vision. *Technology and Engineering Teacher, 70*(1), 30. https://www.proquest.com/scholarly-journals/advancing-stem-education-2020-vision/docview/853062675/se-2?accountid=10207

Bybee, R. W., Taylor, J. A., Gardner, A., Van Scotter, P., Powell, J. C., Westbrook, A., & Landes, N. (2006). The BSCS 5E instructional model: Origins and effectiveness. *Colorado Springs, Co: BSCS, 5,* 88–98.

Hanvey, R. G. (1976). *An attainable global perspective.* The Center for Teaching International Relations, The University of Denver.

Hernandez, J. (2022). *Fresh banana leaves: healing Indigenous landscapes through indigenous science.* North Atlantic Books. https://doi.org/10.1002/lob.10592

Loewen, J. W. (2008). *Lies my teacher told me: Everything your American history textbook got wrong.* The New Press.

Moberg, M. (1996). Crown colony as banana republic: the united fruit company in British Honduras, 1900–1920. *Journal of Latin American Studies, 28*(2), 357–381. https://doi.org/10.1017/S0022216X00013043

Neruda, P., & Schmitt, J. (2000). *Canto general* (50th anniversary pbk. ed.). University of California Press.

Poole, C. M., & Russell III, W. B. (2015). Educating for global perspectives: A study of teacher preparation programs. *Journal of Education, 195*(3), 41–52. http://www.jstor.org/stable/44510416

Postman, N., & Weingartner, C. (1969). *Teaching as a subversive activity: A no-holds-barred assault on outdated teaching methods-with dramatic and practical proposals on how education can be made relevant to today's world.* Delta.

Scherrer, B. D. (2022). "Like you can tell a river where to go": Floods, ecological formations, and storied pedagogies of place. *Curriculum Inquiry, 52*(2), 187–204. https://doi.org/10.1080/03626784.2022.2041977

Tichnor-Wagner, A., Parkhouse, H., Glazier, J., & Cain, J. M. (2019). *Becoming a globally competent teacher*. ASCD. https://files.ascd.org/pdfs/publications/books/Becoming-a-Globally-Competent-Teacher-Sample-Chapters.pdf

White, T., Wymore, A., Dere, A., Hoffman, A., Washburne, J., & Conklin, M. (2017). Integrated interdisciplinary science of the critical zone as a foundational curriculum for addressing issues of environmental sustainability. *Journal of Geoscience Education*, 65(2), 136–145. https://doi.org/10.5408/16-171.1

4
TEACHING FOR RACIAL AND ENVIRONMENTAL JUSTICE IN FLINT, MICHIGAN, USA

Margery Gardner

Introduction

"Every politically engaged person should have a garden," writes poet Camille Dungy (2020, p. 154). She describes how the care for plants in a garden can offer calm amidst times of chaos and worry, like worry for our children's futures and a desire to leave them more than toxic land and limited reproductive rights. Dungy leans on a multitude of identities, including mother, professor, and Black gardener. She advocates for shifts in the discursive ways we talk about nature to encompass a more representational body of experience. Through acknowledgment of place, we can understand ourselves in relationship to our unique identities, immediate and historical communities, and global societies, as well as make new connections to land and water. Dungy locates healing within green spaces and land as life-sustaining. She recognizes that land and water resources are under constant threat and that we must remain vigilant in order to protect them for future generations. Yet, water and land can quickly become an invisible toxic threat. Images of Flint, Michigan's filthy tap water can serve as a visceral allegory for the impacts of environmental racism and community neglect. Flint residents deserve to have their stories heard and remembered.

Case studies can act as powerful tools to teach science content because of their ability to tell compelling stories that relate to common human experiences (Herreid, 2005). The position the teacher adopts is that of a facilitator enabling the learner to analyze the problem at hand or offer future solutions. Since case studies are rooted in the real world, there is an inherent messiness to learning with no predetermined right or wrong answers.

DOI: 10.4324/9781003395782-7

> Use of case studies in science should encourage students to critically appraise stories about science they hear through the media, to have a more positive attitude about science, to understand the process of science and its limitations, and to be able to ask more critical questions during public policy debates.
>
> *(Herreid, 1994, p. 222)*

Learning becomes more process-oriented and focused on critical science literacy development and civic duty. In this chapter, I will discuss ways to teach environmental racism in order to reveal the political motives and transgressions that result in unsafe spaces for Black and Brown children. The environmental calamity that unfolded in Flint, Michigan, over the past decade, which made tap water toxic for its residents, unfortunately elucidates this concept.

The "question worth knowing" for this case study is: Why is the water supply of Flint, Michigan, so toxic? In order to answer this question, learners require an understanding of the feedback, as well as chemical properties, corrosion, and solutions. Mathematically, growth rates and relationships between quantities must also be investigated. When properly defined and supported, models of instruction that weave together subject areas provide an elevated form of instruction that allows learners to engage in authentic tasks that foster cognitive and affective skills (Contant, Bass, Tweed, & Carin, 2017). Students receive a more viable and socio-critical representation of their world. Students who have been exposed to critical and integrated pedagogies are afforded entry points to transfer knowledge (Bransford, Brown, & Cocking, 2000). STEM teaching typically offers little opportunity to understand racism as a structure of power. I argue in this chapter, and throughout the book, that STEM education can and should acknowledge and challenge structural racism through disruptive STEM pedagogies.

Attending to the scientific and mathematical aspects of the Flint water crisis also involves understanding the frameworks that enable environmental racism to persist. Reflecting on David's question posed to learners in Chapter 3, "If someone has done harm then is it possible to undo the harm that you do?" we might add: what if that someone is your own government? Giving learners the apparatus to view intersections of environmental and social issues is a responsibility of all teachers, including STEM educators. Engaging learners in questions regarding why the Flint water crisis occurred and how the government should mitigate the damage is part of a robust STEM learning experience.

Structural and strategic racism and STEM

In the upcoming section, I provide a brief summary of the events in Flint, Michigan, that resulted in severe poisoning of the city's water sources. Professor Terese Olson explained the science behind the Flint water crisis as a "corrosion of pipes, erosion of trust" (2016). The crisis first began in April 2014 as an effort to save money. At that time, Flint, Michigan's water supply was changed from buying treated water from Detroit's supply, which includes Lake Huron and the Detroit River, to the Flint River, known to contain high levels of chloride.

The coagulant (iron salt) used to treat corrosivity in the Flint River was selected based on cost and was not appropriate for the particular water supply. It was not effective at settling the organic materials and actually elevated the amount of chloride in the water. The disinfectant process creates a byproduct (trihalomethanes) that forms from a reaction of chlorine and organic matter. Since there was already a large quantity of chlorine in the water, this resulted in water that was highly corrosive, caused by a positive feedback loop. The iron in pipes reacted with the chlorine, making it ineffective at killing bacteria. More chloride was added to the water to help combat the increases in bacteria. This just perpetuated the problem. Corrosive water, dark in appearance (due to iron deposits) and with a strong odor, began to eat away at the infrastructure. Homes built before 1986 typically contain copper pipes with lead connectors. Pipes from water mains can also contain lead. When corrosion occurred in these older systems, lead and copper were released directly into the water supply. Lead solids were also visible in these water samples. Orthophosphates can be added to water to help chloride ions adhere and reduce the impact of corrosion on pipes. Water management is complex, and, in this case, there was a lack of reliance on scientific evidence to inform decision-making.

Residents began to complain of dirty water running from their faucet, but their concerns were dismissed by governmental entities. Water quality checks by a team of researchers at Virginia Tech found elevated levels of lead from one resident's tap that rose from 105 ppb in January 2015 to 707 ppb by April of that same year. The greatest concentration level recorded was an unfathomable 13,000 ppb. About 200 cases of lead poisoning were confirmed in the first year, with half of exposure out of a modest population of 100,000. To put these levels in perspective, the Environmental Protection Agency (EPA) considers a threshold of up to 15 parts per billion of lead safe for use. The long-term health impacts from this extreme lead exposure are still being assessed, especially in children who may require much greater learning supports. Also linked to improper water treatment was an outbreak of Legionnaire's disease, which killed 12 and caused severe illness in over 75 citizens (Johnson, 2016).

Disruptive STEM teachers can leverage critical facets to tell Flint's story in ways that bring to light factors such as race and class. Also drawing from the interdisciplinary nature of this case, both STEM subjects and sociopolitical dimensions of the Flint water crisis become part of a balanced lesson. The Ambitious Science Teaching model from the University of Washington's School of Education supports this lesson's mission by centering instruction on a perplexing science phenomenon. Learners encounter this phenomenon, referred to as the anchoring event by Windschitl, Thompson, and Braaten (2020). The phenomenon serves as a foundation for subsequent conversations and learner-guided experimentation. The goal is to have learners develop their own ideas about the phenomenon and then seek group consensus for those with the most explanatory power.

In this case, the reported lead levels from the first 18 months of exposure act as the phenomenon of study. One imperative of the Ambitious Science approach is that the science teacher must be well-versed in the causal story that explains the phenomenon at hand and be prepared to guide students through several different engagements with the concept. This approach allows for great student latitude to guide discussions and activities as they seek to amass evidence in support of their scientific claims.

The causal story line in this instance includes both chemical reactions and concentrations, as well as the presence of structural racism. Structural racism is defined as the institutional patterns manipulated by White supremacy to shape outcomes in favor of White, normative, middle class folks. Flint, Michigan, is a majority Black community, with over 40% of residents considered living below the poverty threshold in the United States (Hill, 2021). Hammer (2019) introduces a complimentary concept of strategic racism, also part of this causal story line, defined as the "manipulation of intentional racism, structural racism and unconscious biases for economic or political gain, regardless of whether the actor has express racist intent, although the very act of engaging in strategic racism is itself a form of racist behavior (p. 104)." The intentional disinvestment in Flint's infrastructure under the guise of fiscal austerity is the primary factor involved in the poisoning of these citizens. Strategic racism was exercised through the complicity of governing bodies such as the Karegnondi Water Authority (KWA), the Michigan Department of Treasury, and the Michigan Department of Environmental Quality (DEQ), who chose to use the Flint River as a replacement water source despite its known high levels of corrosivity. This was a blatant act of violence on the people of Flint. By unlocking both STEM and sociopolitical concepts at play in Flint, Michigan, learners are equipped to understand more fully the issues of environmental racism from different angles.

Unfortunately, Flint, Michigan, is just one locality out of countless others impacted by severe political neglect and health uncertainty. Since I first taught this lesson in 2016, there have unfortunately been many other instances in the United States alone. Jackson, Mississippi, garnered media attention recently due to total water outages for some, as well as chronic boiled water notices. Residents report that the water flowing from their faucets is dark brown in color and that they experience skin irritation when showering. The infrastructure of the city includes pipes, installed in 1914, that have repeatedly burst throughout the area. While politicians were aware of the need to remedy this issue, little action was taken. The municipal water department was chronically underfunded and understaffed. One water leak located on the grounds of an old golf course spews 5 million gallons a day, according to the *New York Times* reporter Sarah Fowler (2023, March 22). About 5 million gallons would provide sufficient water for nearly a third of Jackson's current population. In August 2022, President Biden approved an emergency declaration request that allowed the city to access federal assistance. According to Lieff Cabraser's website, the attorneys who legally represent Jackson residents, a federal-class action lawsuit was filed on September 16, 2022, citing that lead levels were above the EPA threshold for safe consumption. In late November 2022, an outside water manager with years of prior experience assumed the lead role to repair infrastructure. Yet, due to years of neglect, extreme pipe leaks persist. Mississippi is ranked last or close to last in the United States for health equity (Mississippi State Department of Health). Due to structural and strategic racism, people of color and those with lower levels of education are most susceptible to chronic diseases such as diabetes and respiratory issues due, in part, to environmental conditions. Water shortage and contamination seek to threaten an already stressed population. Access to safe water is a basic human right that requires our collective attention and advocacy. Unfortunately, environmental justice case studies are still in supply for disruptive STEM educators to confront in their classrooms.

Political pressure to support disinvested communities such as Flint resulted in the creation of a National Office for Environmental Justice and Civil Rights by the EPA in September 2022, with a mission to support historically marginalized, underserved communities. The office is a conglomerate of three others that feature 200 dedicated staff members throughout the United States. The EPA tasks employees in this office with the following:

> These staff will engage with communities with environmental justice concerns to understand their needs, as well as Tribal, state, and local partners; manage and disburse historic levels of grants and technical

assistance; work with other EPA offices to incorporate environmental justice into the agency's programs, policies, and processes, as allowed by law; and ensure EPA funding recipients comply with applicable civil rights laws.

According to the Smithsonian Magazine, the new office plans to disburse 3 billion dollars for projects in communities to combat environmental harms and address the effects of climate change.

While the addition of a new office seems like a progressive step, the EPA established the Office of Environmental Justice (OEJ) over 25 years ago with a similar mission: to promote the rights of all people regardless of race, national origin, or class. The OEJ will be housed under this new office alongside two others. However, reliance on executive orders creates political whiplash, depending on the administration in power. While the OEJ gained momentum during the presidency of Barack Obama, President Trump's widespread disinvestment in the EPA caused significant internal strife. Implementation of environmental justice agendas come without any real consequence for violations, according to Perls (2020) from Harvard Law School. US citizens must remain vigilant to ensure that future incidences of environmental racism are confronted and remediated at the national level rather than being met only with political rhetoric. Disruptive STEM teachers can help learners make sense of political events and give tangible ways to let their voices be heard.

Clean water and land are human rights for all people, regardless of identity. Davis (2019) argues that natural resources are subject to anti-Black policies that limit access through a variety of oppressive mechanisms. Davis implores a fuller examination of this issue, since discourses about Blackness and environment are frequently limited to urban spheres and toxic waste contamination. Admittedly, the curricular centerpiece of this chapter draws in part from this sort of discourse and so should be undergirded with conversations that complicate the topic. Scherrer (2022) emphasizes the need to replace damage-centered narratives (Tuck, 2009) associated with land and water with those that locate and legitimize Black-led and Black-founded environmental justice organizations.

Black survival networks are fortified through a common call to action that has been part of a long-standing means to combat White supremacy. Without the demands for justice by Flint residents, it is uncertain whether any governmental intervention would have occurred. Mari Copeny, also known as "little Miss Flint," was 8 years old at the start of the water contamination. She wrote to President Obama to try to get the attention of national lawmakers. As a result of her advocacy, Obama responded to Mari's letter and visited Flint. Let us heed Camille Dungy's advice and make it possible for us all to have access to life-sustaining natural spaces, to have a garden.

Since environmental problems are "wicked" in scope, there is a need for education frameworks that reflect this complexity. Mathematician, designer, and teacher Horst Rittel was first to use the term "wicked problem," which he defined as a "class of social system problems which are ill-formulated, where the information is confusing, where there are many clients and decision makers with conflicting values, and where the ramifications in the whole system are thoroughly confusing" (Buchanan, 1992, p. 15; Churchman, 1967;Rittel & Webber, 1973). Achieving environmental justice is a wicked problem with global features. Water treatment is complex due to chemical reactions that take place and render substances permanently different. Adding chemicals to the water supply results in unintended consequences to environmental and human health. What we add to water sources impacts our communities in complex ways, with communities of color being the most susceptible to political inaction. Ignoring such issues of environmental justice in secondary-level teaching exacerbates wicked problems and denies learners vital advocacy skills as global citizens. Case studies, such as the Flint water crisis, support citizenship development while encouraging learners to think critically about science and its relationship to social systems and power dynamics. Through a disruptive STEM lens, we can bring into focus these critical and interdisciplinary facets of complex global problems.

As a teen, I participated in collective environmental activism, leading me and my peers to knock on the doors of state and national politicians to demand greater protections for our rural community, which was frequently targeted for landfills, sludge factories, and the similar sites. As I moved into a career in education, I continued this work by inviting youth to take part in advocacy.

The devaluation of certain bodies is unacceptable for a viable and thriving future. As Dr. Bettina Love puts it, "These kids matter." We, as educators, need to advocate and fight to create a more just system that prioritizes environmental well-being. Doing the right thing, even when it is hard, should be embedded in the fabric of all our teaching decisions. We think and act democratically and give learners the tools to do the same. STEM teaching should not conflict with these goals but support, uplift, and empower learners to make informed decisions and dream of a reimagined future. We need the civic participation of our children to help us navigate the potential catastrophic impacts of global climate change and other grand environmental challenges. We need to think and act with our intersectional interests in mind in order to thrive in future episodes of human habitation of our planet.

References

Bransford, J. D., Brown, A. L., & Cocking, R. R. (2000). *How people learn* (Vol. 11). National Academy Press.

Buchanan, R. (1992). Wicked problems in design thinking. *Design Issues, 8*(2), 5–21.

Churchman, W. (December 1967). Wicked Problems, *Management Science, 4*(14), B-141-42.

Contant, T. L., Bass, J. L., Tweed, A. A., & Carin, A. A. (2017). *Teaching science through inquiry-based instruction.* Pearson.

Davis, J. (2019). Black faces, black spaces: Rethinking African American underrepresentation in wildland spaces and outdoor recreation. *Environment and Planning E: Nature and Space, 2*(1), 89–109. https://doi.org/10.1177/2514848618817480

Dungy, C. T. (2020). Reasons for gardens. *Ecotone, 16*(1), 154–158. https://doi.org/10.1353/ect.2020.0029

Environmental Protection Agency (2022, September 24). EPA launches new national office dedicated to advancing environmental justice and civil rights. https://www.epa.gov/newsreleases/epa-launches-new-national-office-dedicated-advancing-environmental-justice-and-civil

Fowler, S. (2023, March 22). A water system so broken that one pipe leaks 5 million gallons a day. *New York Times.* https://www.nytimes.com/2023/03/22/us/jackson-mississippi-water-crisis.html

Hammer, P. J. (2019). The Flint water crisis, the Karegnondi water authority and strategic–structural racism. *Critical Sociology, 45*(1), 103–119. https://doi.org/10.1177/0896920517729193

Herreid, C. F. (1994). Case studies in science-A novel method of science education. *Journal of College Science Teaching, 23,* 221–221. https://static.nsta.org/case_study_docs/resources/Novel_Method.pdf

Herreid, C. F. (2005). *Using case studies to teach science. Education: Classroom methodology.* American Institute of Biological Sciences. https://files.eric.ed.gov/fulltext/ED485982.pdf

Hill, B. (2021, June 14). *Human rights, environmental justice, and climate change: Flint, Michigan.* American Bar Association. https://www.americanbar.org/groups/crsj/publications/human_rights_magazine_home/the-truth-about-science/human-rights-environmental-justice-and-climate-change/

Johnson, G. (2016, December 8). *Virginia Tech researchers explain the Flint water crisis.* University of Notre Dame. https://science.nd.edu/news-and-media/news/virginia-tech-researchers-explain-the-flint-water-crisis/#:~:text=The%20problem%20with%20the%20water,despite%20its%20negative%20health%20effects.perls

Olson, T. (2016, February 4). The science behind the Flint water crisis: Corrosion of pipes, erosion of trust. *The Conversation.* https://theconversation.com/the-science-behind-the-flint-water-crisis-corrosion-of-pipes-erosion-of-trust-53776

Perls, H. (2020, November 12). *EPA undermines its own environmental justice programs.* Environmental and Energy Law Program Harvard Law School. https://eelp.law.harvard.edu/2020/11/epa-undermines-its-own-environmental-justice-programs/

Rittel, H. W. J., & Webber, M. M. (1973). Dilemmas in a general theory of planning. *Policy Sciences, 4*(2), 155–169. https://doi.org/10.1007/BF01405730

Scherrer, B. D. (2022). "Like you can tell a river where to go": Floods, ecological formations, and storied pedagogies of place. *Curriculum Inquiry, 52*(2), 187–204. https://doi.org/10.1080/03626784.2022.2041977

Tuck, E. (2009). Suspending damage: A letter to communities. *Harvard Educational Review, 79*(3), 409–428.

Windschitl, M., Thompson, J., & Braaten, M. (2020). *Ambitious science teaching.* Harvard Education Press.

5

CENTERING INDIGENOUS VOICE

STEM teacher training in service of decolonial futures

Hugh Burnam and Margery Gardner

Introduction

The Haudenosaunee (Hoe-dee-no-SHOW-nee), meaning the People of the Longhouse and also known as the "Iroquois Confederacy," are comprised of six nations in North America, extending across parts of United States in New York and Wisconsin, and also parts of Canada. The unification of these nations, first with five and then extending to six, is considered the earliest form of democracy in North America. The following nations comprise the Haudenosaunee: Mohawk or Kanien'kehá:ka (People of the Flint), Oneida or Onyota'a:ká (People of the Standing Stone), Onondaga or Onoñda'gega' (People of the Hills), Cayuga or Gayogohó:nǫ (People of the Great Swamp), Seneca or Onöndowa'ga:'(People of the Great Hill), and the Tuscarora or Skarù:rę' (The Shirt-Wearing People) (Harris & Johnson, 2009). For the purposes of this chapter, we offer common language, defining terms that are frequently used but rarely interrogated. "Indigenous" or "Native" is a term used to refer to "Native Americans," "American Indians (AI)," "Aboriginal," "First Nations," or Native peoples who live on what some might call "Turtle Island," also known as "North America."

The Haudenosaunee are among the most studied group of Indigenous peoples in North America. Traditional Haudenosaunee teachings influenced the Founding Fathers and the writing of the United States Constitution (Miller, 2015), inspired early suffragists in the fight for women's right to vote (Weller, 2020), and fascinated the likes of Friedrich Engels and Karl Marx through the work of Lewis Henry Morgan and his experiences "among the Iroquois" (Engels, 1942). The Haudenosaunee have been

DOI: 10.4324/9781003395782-8

examined, re-examined, poked, and prodded by Western writers, historians, and politicians for hundreds of years – yet so many non-Native educators, staff, and school administrators who live on Haudenosaunee lands and even neighbor Haudenosaunee communities continue to misrepresent, misunderstand, or completely ignore Haudenosaunee people today. This silencing is evident in curricular offerings in schools. In-depth analysis of texts from the 2011–2012 school year revealed "that nearly 87 percent of state history standards failed to cover Native American history in a post-1900 context, and that 27 states did not specifically name any individual Native Americans in their standards at all." These data suggest that portrayals of Native people in schools tend to historicize or exclude their perspectives altogether (Shears, Knowles, Soden, Castro, 2015).

This chapter centers Indigeneity in STEM education and creates space to contemplate decolonial futures. Hugh shares his work as a teacher trainer and Native scholar to dislodge mindsets of Whiteness in school systems, including the challenges of bringing decolonial perspectives to teachers. Margery uncovers the applications of object-based pedagogy and how its central concepts can be integrated into STEM-based instruction. The work of artisan Ronni-Leigh Goeman brings to light the need for intersectional efforts to promote environmental sustainability from an Indigenous paradigm. The chapter discusses the use of Black Ash trees to create art in the form of baskets and the threats to these trees and associated cultural traditions due to an infestation in the United States of the Emerald Ash Borer. In the latter half of the chapter, we discuss how teachers can design curricula with intentionality that celebrate and center Indigenous people and their many contributions, including stewards of the land. Teachers and teacher candidates put the foundational ideas decolonizing curriculum to practice in various ways based on their own identities.

Decolonization might have different meanings across different communities, but for Indigenous peoples, the act of decolonization has long been rooted in continuing the instructions given to them since time immemorial. For many Indigenous communities, decolonization means sharing oral histories, continuing to learn and speak Indigenous languages, participating in traditional ceremonial practices, and reinforcing land-based relationality. For Indigenous peoples, the land is where people learn to become responsible human beings.

Indigenous peoples often perceive the Western education curriculum as reductive, exclusionary, and dismissive of Indigenous epistemologies and worldviews. Seen through an Indigenous lens, the Western curriculum is a piece of a larger nefarious system intentionally created over the last 500 years to colonize Indigenous lands. This often jarring perception of education curriculum raises an important question that so many educators might grapple with regularly: What would it mean to "decolonize" the curriculum?

Brooke Madden (2015), an early career researcher and teacher educator, describes decolonial educational approaches for educators working within Indigenous education as centered on four pedagogical pathways: learning from Indigenous traditional models of teaching, pedagogy for decolonizing, Indigenous and anti-racist education, and Indigenous and place-based education.

Madden writes:

> Each pathway is differently concerned with the central task of reshaping contemporary Indigenous-non-Indigenous relationships through teacher transformation. Divergent pedagogical methods produce shifts in teachers' positioning and practice that challenge interconnected systems of oppression in schools, and respond to the needs of Indigenous students and communities (p. 12).

Madden's Indigenous education pedagogical pathway offers important insights into a much-needed decolonized approach to teacher education in context of Indigenous studies and with Indigenous students. Similarly, educators in predominantly White schools aiming to decolonize curricula must also continue to prioritize Indigenous-based relationships, despite the perceived absence of Indigenous students and Indigenous communities. Educators are challenged with an often white-washed curriculum designed to dismiss histories of historically marginalized communities, diminish Indigenous knowledge, and remove learners from the land; therefore, they must continue to acknowledge Indigenous lands, work with Indigenous communities, reinforce Indigenous pedagogy and intellect, and continue to examine ways in which Indigenous and non-Indigenous communities are implicated within the larger project of settler colonialism. Disruptive STEM education supports decolonizing efforts through a prevailing critical assumption that pedagogical change is both necessary and possible within secondary schooling spaces. Situating self, one of four facets of disruptive STEM, offers reflective space to understand our own and others social positioning within the contours of a long-term colonial apparatus.

Facing tensions in teacher training

Hugh Burnam Hode'nhyahä:dye' (Mohawk Nation), a co-author of this chapter, engages in work alongside Indigenous co-facilitators to talk with educators about decolonizing their curriculum in schools, colleges, and universities. In addition to working alongside co-author Margery at Colgate University, Hugh has worked with Dr. Lori Quigley (Seneca Nation) and other prominent community educators to bring this work to school districts and into university settings. Some non-Native educators have called this work "cultural competency" or "PD" ("Professional Development"), terms which are reductive of the lived realities of Indigenous, Black, and Brown people within the larger apparatus of settler colonialism. In this chapter, we intentionally refer to this work as "decolonizing the curriculum" in order to center the knowledge and lived experiences of Indigenous peoples and to confront the systems intentionally designed to marginalize them.

In this work, Hugh and his collaborators have found that non-Native educators experience various tensions and conflicts during the trainings, which include the following:

1 Struggling to understand Indigenous sovereignty, self-determination, and Indigenous worldviews
2 Expressing feelings of guilt about their Whiteness, settler colonialism, and fear being called racist
3 Relegating Indigenous peoples to the past and/or downplaying the presence of settler colonialism today
4 Holding deficit-based perceptions of Indigenous peoples while centering themselves as "savior"

Following these four main challenges listed, Hugh offers deeper insight into these challenges and possible ways to address them, as discussed later. He and his co-facilitators often use multiple forms of media, storytelling, and objects to create discussions and dialogue, encouraging educators to ask questions and leaving space for reflection and further discussion. In their work toward understanding the meaning of decolonization within schools, these important areas must be further explored and considered.

Struggling to understand Indigenous sovereignty, self-determination, and Indigenous worldviews

Non-Native educators struggle to understand Indigenous sovereignty, self-determination, and worldviews that are not centered on a mainstream White American life. Indigenous peoples are often perceived by non-Native educators as only being distinguished based on race; but there is much more to the legal and political status of Indigenous peoples. Indigenous peoples in North America have a distinguishable separate nationhood from the United States (and Canada). There are currently 574 federally recognized Indigenous nations or tribes, all of whom operate their own nations through "self-determination" or the ability to govern their own affairs as distinct political entities, recognized by the United States (Nelson, Tachine, & Lopez, 2021).

As facilitators, we have noticed that many non-Native educators do not understand the significance of cultural symbols in schools. To build more awareness around these symbols, facilitators teach about wampum belts and their sociopolitical significance among the Haudenosaunee. Educators are provided with articles, replica wampum belts, and pictures of select wampum belts on slide shows to teach about their significance in establishing political distinction. This approach teaches educators about important historical nation-to-nation agreements, and/or building, refining, or changing the relationships among Indigenous nations or between Indigenous nations and non-Indigenous nations.

The Hiawatha Belt, for example, is a wampum belt which tells the story of the unification of the Haudenosaunee through the "Great Peace" that united the original five nations: Mohawk, Oneida, Onondaga, Cayuga, and Seneca. The belt symbolizes so much rich history about the Haudenosaunee and also serves as a meaningful lesson for educators who can unpack the significance of the use of wampum belts as a lesson in history and governance. Following this, Hugh also uses what is often called the "Haudenosaunee flag." This flag was created to resemble the Hiawatha Belt and is often used to represent or make visible the Haudenosaunee in public spaces. The flag can also be used to symbolize the nation-to-nation relationship between the US and Haudenosaunee people when flown side-by-side (as it is at several schools and institutions) or to cultivate a sense of belonging for Indigenous students at schools. Waterman and Arnold (2010) discuss the importance of cultural symbols in schools and explore the experiences of high school students when Lafayette High School, a New York State public school that serves Native and non-Native students, raised the Haudenosaunee flag. As educators in these decolonizing sessions read this article, they uncover that, on the one hand, Onondaga Nation community members and Indigenous students experienced a sense of pride and celebration when it was raised, but, on the other hand, they experienced pushback, debate, and controversy from members of the neighboring conservative community.

Learning about wampum belts and cultural symbols in schools forces people to shift their worldview from a Euro-American framework to an Indigenous framework, a task that may be difficult for many American educators. The potential controversy that follows raising flags as cultural symbols might seem trivial, or even silly, but across differences in worldview, culture, and political understanding, this distinction is quite significant. Teaching about wampum belts, their histories, and their use today can inform educators that Indigenous peoples are not simply a racial group, but legally recognized as distinct Indigenous nations with a rich history.

Expressing feelings of guilt about their Whiteness, settler colonialism, and fear being called racist

Non-Native educators who identify racially as White may struggle with their feelings of guilt associated with their racial identity, how they are situated within the context of colonization, and fear of being called racist. Educators often struggle with understanding systemic ways in which racism and settler colonialism operate, not only in overt ways, but most prominently in subtle ways that perpetuate systems of inequity all around us and especially in schools. Failure to understand the complexities of systemic inequity exacerbates the conflicted and confused feelings that educators might have about race and colonization in their everyday lives. We remind educators that our approach to this work is "not to call out, but to call in" which extends a sense of welcome to uncomfortable or challenging conversations.

Facilitators ask educators what kinds of privileges they might have due to ongoing colonization; examples could include privilege of race, citizenship, wealth acquisition, and land occupation (Joyce, 2022). Facilitators also ask participants to reflect on their family's arrival in the Americas or about lands that they're on. It is useful to ask if educators can name the Indigenous lands that they currently work on or reside on. Some educators may become silent throughout the conversation.

The facilitators then ask educators to think and talk about how they address systemic racism in schools and to reflect on the idea of settler colonialism. In order to teach about systemic inequities, the facilitators provide words or phrases to educators, including "systemic racism," "racial microaggressions," "stereotypes," and "multi-generational trauma," to name a few. Without revealing the definitions, facilitators ask the educators in the room to unpack the definitions of each of the words or phrases in their own words. As educators make attempts to define each, it is common for only a few educators to fully and actively participate in the exercise, while others in the room more passively observe and the space becomes quiet.

After educators attempt to unpack each word or phrase, the facilitators reveal a full definition of each and how each word or phrase applies to Indigenous education or Indigenous experiences in schools. For example, when facilitators reveal "white washing," they are provided with settler myths like "Christopher Columbus discovered America" and asked to respond. In most cases, educators agree that this is a common settler myth in schools that white-washes history. Hugh and the facilitators express that white washing in schools is a prominent form of systemic racism, and an everyday occurrence. White educators may voice a perceived sense of guilt surrounding racism and settler colonialism and may become emotional or express sentiments like, "that's not my fault" or "I'm not racist." In some cases, educators want to move on from the topic or express the need for a "safe space" in an effort to distance themselves from feeling guilt associated with systemic inequities.

In response, facilitators provide educators the opportunity to create their own "guidelines" for the space. After asking each participant to provide their own guidelines, educators most commonly offer examples like "agree to disagree," "make eye contact," and "maintain a safe space." Hugh then provides educators with the article, *From Safe Spaces to Brave Spaces* (Arao & Clemens, 2013), to unpack suggestions they might have made for the space. The article explicitly reveals concerns around sentiments like "agree to disagree," for example. While educators often offer "agree to disagree" as a solution for conflict in the space, Arao and Clemens (2013) write, "we believe that agreeing to disagree can be used to retreat from conflict in an attempt to avoid discomfort and the potential for damaged relationships" (p. 39).

Another prominent example is educators wanting to create a "safe space," which is understandable and actually an important recommendation for schools in particular. However, Arao and Clemens (2013) describe the limitations of using "safe space" when engaged in social justice pedagogy and dialogue, noting that perceptions of "safety" within this context can be expressed when group members face factors associated with their own unearned privilege. When they hear difficult stories from those experiencing systemic inequity, educators in the group experience "direct challenges to their worldview... which elicit a range of negative emotions such as fear, sorrow, and anger" (p. 139). As educators read this article, they often become reflexive about their own positionality and, in most cases, begin to understand that "safe space" is limited. Hugh and facilitators suggest, following the article, that educators need to cultivate a sense of bravery and courageousness in order to benefit from the space – which can be an empowering exercise for most.

Relegating Indigenous peoples to the past and/or downplaying the presence of settler colonialism today

Non-Native educators might struggle with identifying settler colonialism, not only as an issue of the past but also as an ongoing issue today. In trainings, Hugh and co-facilitators usually screen *Unseen Tears: The Native American Residential Boarding School Experience in Western New York* (Douglas, 2009), a documentary that shares stories from survivors of residential boarding schools that were located in Western New York (Thomas Indian Boarding School) and Ontario (The Mohawk Institute, aka "The Mush Hole"). The documentary reveals ways that Native families were and still are impacted by the lasting generational traumas inflicted upon them by state-sponsored efforts to assimilate Indigenous children to American life. The schools were in operation from the late 1800s well into the 1980s. Survivors told their stories during the documentary's filming in 2008. When we stream this documentary in the classroom for educators, they may express surprise and state that the film is "shocking" to them. While this may be "shocking" for many people to learn, for so many Indigenous peoples, their families and communities, this history still reverberates in everyday life today.

Streaming this documentary might also lead to challenges. When educators express shock and relegate these atrocities to the past, they also fail to connect ways that similar instances happen in Indigenous people's lives today. In dialogue, educators emotionally asked, *"How could this ever happen?"* in order to demonstrate their disapproval of residential boarding schools. But in the same dialogic space, such educators openly described that, despite Indigenous student sovereignty, it is important for Indigenous students to stand for the pledge of allegiance to "respect" the US flag, these educators ask for "simpler ways" to say Indigenous students' names, or they assume that Indigenous students more commonly come from broken homes or that they are "poor" and always in need. These are only some examples of teachers' deficit-based attitudes about Indigenous students that shape how they teach Indigenous-based content to classes of predominantly non-Native, White students.

Facilitators bring in guest speakers and share videos and written stories from Indigenous authors and other content from contemporary Indigenous life to draw connections with the past atrocities brought on by residential boarding schools. Educators often need to meet Indigenous peoples, who might bring in their scholarly work, their art, their traditional knowledge, and their personal experiences to describe ways that settler colonialism is still present every single day. This approach helps educators connect past events with the experiences of today's Indigenous peoples. We have specifically highlighted issues of trauma that reverberate from the past and are enacted today but also emphasize Indigenous strengths, lifeways, resiliency, and survival. This is deeply connected to the next section.

Holding deficit-based perceptions of Indigenous peoples while centering themselves as "savior"

When learning about Indigenous culture, history, and experiences, educators in the space often grapple with their perceptions of Indigenous peoples, especially deficit-based assumptions. Facilitators have heard, "But I just want to help," and while helping is certainly an important part of the work, there is a fine line between "helping" and centering themselves as a hero or savior to Indigenous peoples. The following description might assist non-Native educators who want to learn about ways to move from a deficit-based to an asset-based perception; and from a savior position to allyship.

We show educators a historical document that was used in an attempt to insidiously steal Indigenous lands in New York State, which predominantly draws from a collection of testimonies from non-Native educators, church leaders, and college or university officials. The Report of the Special Committee to Investigate the Indian Problem of the State of New York, also known as the Whipple Report (1889), investigated what they called "reservation life" among the Indians of New York State in 1888 through almost solely a deficit-based lens. This controversial report would generate letters to Chairman Whipple to inform legislation in New York State with a strong emphasis on full Haudenosaunee assimilation into US society and a suggestion that their lands be given to the United States. Full assimilation and full integration were championed by education administrators, including multiple school superintendents in Upstate New York and Charles Sims, then Chancellor of Syracuse University (Whipple, 1889). This controversial report justified the large-scale assimilation of Haudenosaunee children into residential schools in New York State. Through use of this document, facilitators warn about the issue of a large-scale deficit-based lens, racialized assumptions, and internalization of White saviorism.

Schools may rely too heavily on deficit-based models of Native student persistence, focusing too often on poverty, low academic performance, violence, or abuse, and not enough on strengths-based models like language revitalization, cultural integrity, Native community, and self-determination, creating more school/campus hostility for students to overcome (Waterman, Lowe, & Shotton, 2018). Deficit-based models are heavily influenced by cultural and racial bias of people in positions of power. This may also lead to de-centering BIPOC students and centering White people who are "helping" BIPOC, moving again toward a "White savior" model. Deficit-based and White savior models lead to a lack of trust, lack of sincerity, and perpetuate systems of inequity. These approaches negatively affect student confidence, self-esteem, and mental health.

Allyship for educators might mean to listen, read, and do their own research outside of what is required in an Indigenous "PD" at their workplace. This action, to us as facilitators and as Indigenous peoples, is truly our goal. We ask non-Native educators what they can do to intentionally de-center themselves and center Indigenous peoples in the curriculum, and by extension, become "allied" to Indigenous communities to empower Indigenous communities and to empower themselves in the process. We ask all educators: How can we decolonize the curriculum?

Centering cultural connections to land

In Steve Wall's (2002) *To Become a Human Being*, Leon Shenandoah, an Onondaga Chief who held the title of "Tadodaho" said, "Another instruction is to give thanks to Mother Earth, as well. Giving thanks is to give honor and to honor is to show respect. Doing that means you're becoming a Human Being" (p. 40). Students who learn the New York State curriculum often spend most of their time within the confines of the classroom learning rigid interpretations of history and do not often get a chance to engage with the land on which they are situated. Land-based relationality as a decolonial praxis aims to reinforce the deep acknowledgment, gratitude, and sense of belonging on the land as human beings. This land-based relationship is explained further through the example of the Black Ash.

Black Ash is the primary tree species used to make baskets by Haudenosaunee and other First Nation people. It serves as an irreplaceable cultural resource that becomes central to the tradition of cutting, processing, and eventually constructing baskets. By the early 1990s, Black Ash in the southeast Michigan area was under serious attack by the Emerald Ash Borer. The Emerald Ash Borer is an Asian beetle that is about half an inch in length with a metallic green coloration. It looks small and slender when compared to other endemic species such as the Tiger Beetle.

The Emerald Ash Borer carries out its life cycle over a 1- to 2-year period that includes overwintering in the bark of its unfortunate host. Adults emerge in late spring or early summer and reside until August, when they lay their eggs. Larvae feast on the phloem and cambium of the tree, leaving snake-like scars on the outside of the sapwood bark. These larvae eat and grow from mid-summer to fall at the expense of the host tree. In parts of Asia, such as China, where the Emerald Ash Borer is endemic, the insect targets compromised trees. However, in the United States and Canada, the beetle infests perfectly healthy ash trees, girdles them, and rids them of their canopy that is essential for photosynthesis. White, Green, and Black Ash are heavily impacted by this invasive species. The Emerald Ash Borer also doesn't seem to delineate by habitat. Ash trees found in swampy areas, like Black Ash, are impacted just as much as trees found in urban areas (Herms & McCullough, 2014).

Since the Emerald Ash Borer was officially identified as a harmful invasive pest in 2002, the beetle has decimated stands of trees in southeast Michigan. By 2010, extreme tree fatalities were reported, with 99% of all ashes greater or equal to 2.5 cm in diameter completely wiped out in the area. Ash trees throughout the Northeast are also feeling the strain of this species. Herms and McCullough (2014) report, "An analysis of the economic impacts of nonnative forest insects found that EAB is already the most destructive and costly forest insect to invade the United States" (p. 17). Much of this cost includes removal of dead or dying trees as well an all-out embargo of ash trees from nurseries in the area. The ecological impacts of this form of devastation are also immense, creating unforeseen consequences on forest and riparian health. Mature trees have also supported humankind through their absorption of toxins and excess water as well as mental health benefits in addition to the cultural contributions through the art of basketmaking.

The work to conserve remaining ash trees and protect future populations is an ongoing struggle. The spread of the Emerald Ash Borer is projected at 20 km per year. Educational campaigns can support the reduction of this spread by cautioning against the transport of ash trees in the form of firewood. Remediation efforts by a number of interested parties including tribal nations have been put into action to preserve the genetics of our ash trees through seed collections. Those individuals that seem to maintain natural resistance to the Emerald Ash Borer are being studied to learn more. Also, Asian Ash trees, where the Emerald Ash Borer acts as a secondary colonizer, are investigated to learn more about the genetics associated with co-evolution. While human-driven dispersal seems to have straightforward actions to prevent the spread, scientists are unsure of the mechanisms for natural movement.

Ronni-Leigh, a highly talented basket-maker and Onondaga eel clan member whose name is Guynehgwenta (which translates to "she knocks down deep snow"), explains the severity of Black Ash loss on cultural sustainability:

Once the Emerald bore is no longer here... they can do a reforestation, the problem is there's not going to be basket makers, it's such a process... There's not going to be basket makers to pass on that tradition. And what I'm hoping is some young people can say, well, 'I remember this' and another one says, 'I remember that' because it's going to be like a 20–30 year process to do the reforestation... we have to wait for the borer [to be] completely gone, reseed it, and let the three get big enough so you can actually pound it and start the process of making baskets, so that's where we stand now.

(Interview, 3/21)

She stresses the importance of teaching children how to make baskets in the hopes that the art form can be viable for the future. She goes on to describe a sense of personal responsibility she feels to carry on the tradition of basketmaking, recognizing it's just one of many native traditions that are facing permanent loss. If the tree stands are decimated as they are trending now, it may take upwards of 30 years for them to recover if intensive conservation efforts are carried out. This delay has serious generational impacts for Native people considering that many of the highly skilled basket makers of today are getting older.

In Kimmerer's (2013) book, *Braiding Sweetgrass*, she pushes back on colonized ways of thinking about ecological interdependence. She critiques the terms "sustainability" and "management" as Eurocentric and focused on human advantage, mainly economical. "Restoring land for production of natural resources is not the same as renewal of land as cultural identity" (p. 328). Situating biodiversity loss as also a loss of Indigenous cultural identity may not be a concept encountered extensively at the secondary level, yet it is certainly worthy of interrogation.

Among educators, there is a desire to shine a more prominent, decolonial light on Native history and culture but a hesitancy by settlers to locate an appropriate path forward. In order to address this need in a way that could actualize a critical curricular opening, we collaborated with people at Colgate University museums, some who are members of the Onyota'a:ká or Oneida Nation. It was our hope that by dedicating class time and space, we could design curated materials that could circulate locally to support teachers wanting to embrace a more honest perspective about Haudenosaunee history and culture. Over a 2-year period, teacher candidates constructed curricula that sought to disrupt some of the antiquated ways that Haudenosaunee history and culture is depicted. The curricular project also challenged teacher candidates to forge connections between people, land, and environment, namely through the centering of Haudenosaunee objects or belongings as means of illustration. We landed on baskets as a belonging that is representative of rich cultural practice and the creation of which is dependent on healthy woodlot ecologies. Lesson plans featured Black Ash baskets as a means to share developing knowledge of worldly interconnectedness.

In August of 2021, Margery hosted an event for student teachers, their cooperators, and other interested teacher partners that centered Indigenous local knowledge, art, and the environment. We began the session with contextualization, using Robin Wall Kimmerer's *Braiding Sweetgrass* (2013) chapter entitled "Wisgaak Gokpenagen: a Black Ash Basket." The group discussed the chapter and related its contents to their own experiences regionally. One participant, an alumnus of Colgate and a current Earth Science/Biology teacher, recounted how she had processed a Black Ash and created splints by hand. Although this experience had occurred years prior, she shared how she continued to keep these splints even if she didn't have any immediate plans to use them because of the amount of labor involved in obtaining them. Other cooperators also made connections about connectedness and how they used maple trees to generate sap on a small scale on their own woodlot and how they made a point to share this activity with their children in hopes of passing along an ethic of care for the natural world.

During the afternoon, the group visited the Longyear Anthropology Museum located on the Colgate campus, where they were able to see up close baskets made by local Haudenosaunee artisans. Vast contrasts between the baskets based on functionality were made very evident through this examination. Student teachers, cooperators, and alumni were able to appreciate the level of intricacy of the woven "fancy" baskets displayed. This session brought to life some of the imagery from Kimmerer's chapter and allowed these educators an opportunity to think through possibilities of connections in their own teaching. For instance, one science teacher alum noted that the ash trees at their school were impacted by the Emerald Ash Borer. She thought that there could be an opportunity to have students conduct some inquiry in their own school yard that could even lead to future advocacy. Duhs (2011) reinforces the need to design experiences that are scaffolded both before and after viewing. Participants should be provided time and space to interrogate their own situated identities and pose questions to the group both during and after viewing time (Figure 5.1).

Across decades and through waves of colonization, museums in higher education in Westernized institutions acquired massive collections that

FIGURE 5.1   Teachers and teacher candidates exploring baskets.

transformed into a physical curriculum for faculty and students of higher education. Object-based learning or OBL was considered a "niche pedagogy" for higher education in part because of its extensive use of university collections in the 18th and 19th centuries. Since that time, museums continue on a path of reimagination to encourage open inquiry that appeals to the senses and fosters productive dialogue about art/cultural belongings (Barlow, 2017).

OBL has made a significant pedagogical comeback, but its incorporation must be tempered by compliance with human rights legislation. The Native American Graves Protection and Repatriation Act or NAGPRA is a US federal law first enacted in 1990 that seeks to protect and restore cultural belongings as well as ancestral remains. NAGRPA also provides guidelines for museums, federal agencies, and institutions of higher education to return items to lineal descendants (National Park Service, 2024). Under the latest set of regulations, free, prior, and informed consent must be received from lineal descendants, culturally affiliated tribal nations, or Native Hawaiian Organizations before the exhibition of, access to, or research on ancestral remains or cultural belongings. Compliance is required for all institutions receiving federal funding (US Department of the Interior, 2023).

From a theoretical standpoint, OBL shares commonality with problem-based and experiential learning. OBL gained traction as a well-substantiated approach that allows for learning to be a fully sensory experience. The outcomes of OBL can range from sheer inspiration to enhancement of cross-disciplinary understanding. There is an affective attraction to objects that binds such forms of learning. By taking advantage of our natural propensity to see, touch, smell, and even at times taste and hear, OBL makes more durable connections with the content, which can assist in longer-term memory recall and application. OBL has been lauded for supporting cognitive engagement with troublesome knowledge because of the use of reflective pauses that enable the learner to contend with difficult concepts, sociopolitical influences, or even emotional responses to injustices such as racism. Participants extract meaning through a more embodied relationship with the content and concepts that situates the educator as a facilitator rather than a directional driver of knowledge. Chatterjee (2011) explains that OBL can be an effective means to foster "discussion, group work, and lateral thinking" (p. 179).

The use of physical materials at the secondary level of education warrants consideration because of its power to attract and maintain attention. We see many different ways in which science teachers, in particular, could bring in more sensory engagement as a means to pique interest and guide discussion that allow for our whole selves to be invested in the learning process based on sensory awareness and not just cognition.

Haudenosaunee representation in curriculum

After attending the session, a teacher candidate quieted the room with this question: "How could White teachers talk about Haudenosaunee history and culture with predominantly White students without perpetuating Whiteness either intentionally or inadvertently?" This comment acted as a driver for this chapter and a reflection point for curriculum development work moving forward. It speaks to the arguments that Hugh and his colleagues make when holding Indigenous-focused professional development sessions. In alignment with Hugh's work with in-service teachers, there was hesitancy among secondary-level non-Native teacher candidates to engage with Haudenosaunee teachings, out of fear that it might be perceived as fraudulent or even as cultural appropriation. This seems to be a healthy tension that these teacher candidates contended with while making curricular decisions.

We worked to confront this question at critical junctures and to extend tools to students with which they can also think about identity and power. This student opted to directly teach students about colonization through dialogue about terms such as "unceded," meaning not handed over or unyielded land. By unpacking this term and posing the question, "Whose land are we on?" to students during introductory lessons, she was able to give students the language to expose more complexities about coloniality, which are often depicted in the past tense rather than being active measures of oppression. It is our intention to provide White settler teacher candidates with resources and knowledge to work toward allyship with their Indigenous communities.

Storytelling as a window to understand cultural beliefs and values was made visible in the curricular materials the teacher candidates created. One teacher candidate used stories of morality from the Oneida Nation, such as the *Legend of How the Bear Lost His Tail*, juxtaposed with European fairy tales to give students an opportunity to analyze and find both difference and commonality between the literary works. Through this comparison, she worked to construct a pedagogical bridge that offered opportunities for text-to-text and text-to-self revelations. Another teacher candidate pushed this notion further to acknowledge that history can be guilty of selective storytelling that silences marginalized voices. He featured *The Peacemaker Story* and shared the narrative of democratic ideals. He drew a critical thread that equipped students to compare the democratic goals of the United States with Haudenosaunee forms of governance. The lesson posed the question to the class, "What have the history books left out when telling the story of founding America?" Drawing on big picture questions vital to Postman and Weingartner's (1969) vision of subversive teaching, critical substance emerges for secondary students to grip onto

FIGURE 5.2   The Legend of How the Bear Lost His Tail cover. Image courtesy of the Oneida Indian Nation.

with space to come to their own conclusions about complex social conditions (Figure 5.2).

One teacher candidate was heavily influenced by the work of local artist Ronni-Leigh Goeman. Students explored the economics behind basketry and the origins of "pretty baskets" as a means to obtain household income. An example of one of these pieces of art is displayed later, symbolically representing the creation story. The New York State K-8 Social Studies Framework includes a section on *Economics and Economic Systems*, where students explain how scarcity necessitates decision-making and employ examples from the Western Hemisphere to illustrate the role of scarcity historically and in current events. This framework created a basis for discussing the decisions by artisans to market and price their creations. Given our capitalist economy, this curricular approach encouraged students to recognize costs associated with labor and materials, in comparison to household needs. Using computational thinking, students at the middle school level were expected to calculate the price points for baskets that adequately reflected effort expended. This could be completed as a purely contemporary task or

FIGURE 5.3   Creation story basket. Photograph courtesy Picker Art Gallery, Colgate University.

one that is part of a historical inquiry that investigated cost of living associated with particular eras (Figure 5.3).

Ronni-Leigh also discussed the evolution of basketry as a utilitarian endeavor. For instance, there was a need for baskets for washing corn where the weave differed depending on the food stuff, such as a double weave for corn meal or soup. Baskets are part of many ceremonies as well, including weddings where they are traditionally exchanged by bride and groom. Fancy baskets, delicate and ornate with threads of bright color and intricate shapes, originated as a means to obtain financial stability for artists who sold them to tourists or collectors. She elaborates that decades prior, making a living as a basket-maker was a serious struggle because prices were too low in comparison to the amount of labor involved. Ronni-Leigh vowed when she began her business, "I'm not going to give away my baskets, I'm not going to give away my time." She collaborates with her husband, who is a master sculptor, to include renditions of bones of animals and people that enhance the already breathtaking pieces (Interview, 3/21).

This teacher candidate rounded out the lesson with narratives that celebrated the artistry and resilience of Native peoples rather than centering conversations around pain and trauma. The lesson gives students different methods of engagement with the concepts, including one performance-based task where students are asked to write about colonization and climate change as confounding threats to Haudenosaunee traditions. One portion of a lesson centered on the Emerald Ash Borer and included discussion of what happens if Black Ash is no longer available for use. Classroom discussion extends to alternatives and next steps. The lesson also looks into collaborative conservation efforts between states and nations to protect Black Ash. There is a clear sense of solidarity in the way the questions are posed to students followed by a call to action. The student teacher asks the class to write letters to prominent basket makers to gather a sense of how they can help support the cause. This empowerment exercise worked to foster a sense of collective urgency and allowed students to see themselves as change makers in their own communities with regard to environmental issues.

Disruptive STEM connections

This chapter provides a theoretical basis or access point for decolonization of curriculum that actively confronts the tokenizing of Native people. Criticality in how we use language and visual representation of Native people and their belongings is part of continued commitment to allyship.

Reexamining the subject structures in place helps us to fully realize the bountiful connections that are ever present as we learn from and continue to make new histories. It is impossible to disentangle the legacies of coloniality from our current environmental predicaments, which we will unfortunately pass along for our children to contend with. The story of cultural continuity intertwines fully with the story of environmental resilience of the Black Ash. Furthermore, objects like Black Ash baskets can serve as power texts that students can read to understand cultural and environmental connection.

Subversion is conceptualized in this chapter as the bravery to develop a more contextualized understanding of species extinction by tracing a line back to settler colonialism. This subversion requires humility as settlers to understand our own complicity in a system that continues to harm Native people and disenfranchise them from their land. The use of object-based learning supports counter narratives that transfer decolonial messages to students with more dimensions.

The self is acknowledged here in ways that definitely feel uncomfortable but, even more so, raw and unnerving. White settler educator instincts to pull back from these conversations must be tempered in order to provoke new more inclusive outlooks.

Disruptive STEM teaching, as defined in this book, centers on the critical examination of intersectional identity of both teacher and learner through pedagogical performances that are subversive, interdisciplinary, and reflective.

This chapter presses us all to reach our growing edges of understanding of how coloniality operates in opposition to native stories and respect for land and Indigenous knowledge.

Final thoughts

In November of 2022, the Colgate University repatriated over 1,500 objects to the Oneida Nation, including funerary objects that were acquired in the 1950s and housed at the Longyear Museum of Anthropology. This was the largest single transfer repatriation effort in New York State history.

Oneida Indian Nation Representative Halbritter responded on the Nation's website (2022):

> For decades, too many museums and other educational and cultural institutions have followed indefensible practices regarding the ancestral remains and cultural artifacts of Native Americans. These practices have been allowed to continue under the belief that preserving history is of the ultimate importance without questioning the means of doing so. They assume it is possible while divorcing the history from the people to whom it belongs, presuming to tell our stories with stolen artifacts and unfamiliar voices. Native people's funerary and ceremonial objects should never be the property of museums in this way. Our dedication to continuing this conversation is one of the many values the Oneida Indian Nation shares with Colgate University. We are grateful for these efforts, but equally grateful for the university's and museum's understanding that they are what is required in a society that meaningfully recognizes the sovereignty and dignity of Native people.
>
> *(Press release, November 10, 2022)*

This quote surfaces the deep and unyielding connection we have to belongings as keepers of culture. Using physical materials requires care as well as the responsibility to honor the whole truth. We pose the question to educators again: What steps can you take to decolonize STEM curriculum?

References

Arao, B., & Clemens, K. (2013). From safe spaces to brave spaces. *The art of effective facilitation: Reflections from social justice educators*, 135 (p. 150). https://www.anselm.edu/sites/default/files/Documents/Center%20for%20Teaching%20Excellence/From%20Safe%20Spaces%20to%20Brave%20Spaces.pdf

Barlow, A. (2017). Beyond object lessons: Object-based learning in the academic library. In *The experiential library* (pp. 27–42). Chandos Publishing. https://doi.org/10.1016/B978-0-08-100775-4.00003-0

Chatterjee, H. J. (2011). *Object-based learning in higher education: The pedagogical power of museums.* https://doi.org/10.18452/8697

Douglas, R. (2009). *Unseen tears: The Native American Boarding School experience in Western New York.* [Film]. Native American Community Services of Erie and Niagara Counties. https://search.worldcat.org/zh-cn/title/613983241

Duhs, R. (2011). *Learning from university museums and collections in higher education.* University College London (UCL). https://doi.org/10.18452/8698

Engels, F. (1942). *The origin of the family, private property and the state.* International Publishers.

Harris, A., & Johnson, T. (2009). *Haudenosaunee Guide for Educators.* National Museum of the American Indian, Smithsonian Institution.

Herms, D. A., & McCullough, D. G. (2014). Emerald ash borer invasion of North America: History, biology, ecology, impacts, and management. *Annual Review of Entomology, 59*(1), 13–30. https://doi.org/10.1146/annurev-ento-011613-162051

Joyce, S. A. (2022). *11: Am I a settler?. White benevolence: Racism and colonial violence in the helping professions.* Fernwood Publishing.

Kimmerer, R. (2013). *Braiding sweetgrass: Indigenous wisdom, scientific knowledge and the teachings of plants.* Milkweed Editions.

Madden, B. (2015). Pedagogical pathways for Indigenous education with/in teacher education. *Teaching and Teacher Education, 51,* 1–15. https://doi.org/10.1016/j.tate.2015.05.005

Miller, R. J. (2015). American Indian constitutions and their influence on the United States Constitution. *Proceedings of the American Philosophical Society, 159*(1), 32–56. http://www.jstor.org/stable/24640169

National Park Service (2024, January 12). Native American graves Protection and Repatriation Act. Retrieved from https://www.nps.gov/subjects/nagpra/compliance.htm.

Nelson, C. A., Tachine, A. R., & Lopez, J. D. (2021). Recognize it's our land, and honor the treaties *Changing the Narrative, 15.* https://www.luminafoundation.org/wp-content/uploads/2021/02/borrowers-of-color-2.pdf#page=15

New York State Legislature Assembly, Special Committee to Investigate the Indian Problem, & Whipple, J. S. (1889). *Report of special committee to investigate the Indian problem of the State of New York: Appointed by the Assembly of 1888. Transmitted to the Legislature February 1, 1889.* Troy Press Company.

Oneida Indian Nation. (2022, November 10). Press release-Colgate University repatriates more than 1,500 funerary objects and to the Oneida Indian Nation, apologizes for acquisition of cultural artifacts. https://www.oneidaindiannation.com/colgate-university-repatriates-more-than-1500-funerary-objects-and-to-the-oneida-indian-nation-apologizes-for-acquisition-of-cultural-artifacts/

Postman, N., & Weingartner, C. (1969). *Teaching as a subversive activity: A no-holds-barred assault on outdated teaching methods-with dramatic and practical proposals on how education can be made relevant to today's world.* Delta.

Shear, S. B., Knowles, R. T., Soden, G. J., & Castro, A. J. (2015). Manifesting destiny: Re/presentations of indigenous peoples in K–12 US history standards. *Theory & Research in Social Education, 43*(1), 68–101.

United States Department of the Interior. (December 6, 2023). Interior department announces final rule for implementation of the Native American Graves Protection and Repatriation Act. https://www.doi.gov/pressreleases/interior-department-announces-final-rule-implementation-native-american-graves

Wall, S. (2002). *To become a human being: The message of Tadodaho Chief Leon Shenandooha,* Hampton Roads Publishing Company, Inc.

Waterman, S. J., & Arnold, P. P. (2010). The Haudenosaunee flag raising: Cultural symbols and intercultural contact. *Journal of American Indian Education*, 125–144. http://www.jstor.org/stable/43608593

Waterman, S. J., & Lowe, S. C., Shotton, H.J. (2018). *Beyond access: Indigenizing programs for Native American student success*. Stylus Publishing, LLC. https://doi.org/10.4324/9781003443230

Weller, K. (2020). Teach NY: Exploring Haudenosaunee influence on America's suffrage movement with students. *New York History*, *101*(2), 377–382. https://muse.jhu.edu/article/775542/pdf

Whipple, J.S. (1889). Special Committee to Investigate the Indian Problem, New York State Legislature. (Assembly Document No. 51, 1889) E78 N7 N77 1889a.

THEME 3

Process-focused, disruptive STEM pedagogy, with interwoven thematic explorations

6

QUEERING THE BIOLOGY CLASSROOM

Intersections between queer pedagogy and disruptive STEM teaching

Providence Rubey

Introduction

The application of critical pedagogy in the classroom faces significant challenges in the United States. As of May 2023, 471 anti-LGBTQIA+ bills targeting schools and education have been introduced across 45 states (ACLU, 2023). Only two months earlier, in March 2023, that number was only 271 bills. In 2021 and 2022, 308 legislative efforts against critical race theory and anti-racism in schools were introduced across the nation (Alexander, Baldwin Clark, Reinhard, & Zatz, 2023). These efforts to dissuade and punish teachers for engaging students in discussions of identity in the classroom have been ramping up in recent years, with no signs of slowing in the near future. In the face of these bills, teachers across the nation are struggling to understand what repercussions they may face for teaching through a critical lens.

This is the backdrop against which I found myself beginning to write this chapter, reflecting on disruptive STEM teaching, what that means to me, and how I see myself as a disruptive STEM educator. Reflected in the forthcoming chapter are the ways in which my identities as a White, queer person, and queer pedagogue impact my teaching positionality. I spent the first six years of my career in the New York City Department of Education as a biology teacher. Now, in year 8 as a biology and earth science teacher in Phoenix, Arizona, I find myself understanding disruptive and subversive STEM pedagogy in very different ways. My experiences and reflections have led me to the conclusion that, ultimately, my job as a science pedagogue is to create a place of criticality and inquiry in the classroom, that encourage

DOI: 10.4324/9781003395782-10

students to ask and answer major questions about themselves and the world around them through the lens of biology. Throughout my time in the classroom, and now more than ever, I see my purpose as an educator to inspire criticality in my students.

I first saw this modeled in my own educational experiences by my high-school biology teacher. While I didn't know it at the time, her teaching practices would lay the foundation for my approach to STEM teaching years later. Her pedagogy situated students as leaders in the science classroom. She used biological concepts as conduits through which we learned to question the world around us, and she taught us to use our critical thinking and investigation skills to seek out our own answers to those questions. We would often engage in inquiry-based labs in which we were given raw materials, with which we designed and implemented the lab work based on what we were most interested in.

This emphasis on critical thinking and question-asking empowered me to connect my learning to my lived experiences. One example of the impact of this teaching approach on learning occurred in my senior year of high school. In my IB Biology class, we were learning about the enzyme catalase, which is found in organisms' livers and breaks down excess hydrogen peroxide in the body. After learning about this enzyme, we were given the task of selecting an "out-of-the-box" independent variable that we could test to see if it affected the speed at which catalase did its job. We could choose anything we wanted, as long as we could write the methods to implement our experiment. This felt freeing to me. I had dozens of scientific questions I wanted to answer, and the fact that I got to be in the driver's seat, choosing what to test and why, empowered me to investigate biological questions that were also deeply personal. I decided I was going to examine the impact of alcohol-based cold medicine on the functionality of the catalase enzyme in the liver. Growing up, adults that I loved struggled with alcohol use, and I was invested in understanding how alcohol might affect their bodies. I designed this experiment from beginning to end, collected data, and completed a final report of my findings. Having flexibility, creativity, and criticality at the center of this biology course empowered me to pursue something in this experiment (while maybe not perfectly scientifically sound!) that would get at a question that gnawed at me, both personally and scientifically.

In my experience as a biology student, I was lucky to encounter a teacher whose goal was not just to have us memorize and regurgitate science facts, but to use those facts as foundations to pursue bigger, broader, and rigorous questions about the world. This experience as a student was the first time I saw school as a place that was meant to be thought-provoking and liberating. As a biology teacher, this is what I strive to recreate in my classroom. And let's be honest, creating this experiential and critical learning space can be challenging! As educators, we often find ourselves in overpopulated and under-resourced classrooms, making it hard to access the tools we may want and need to push our students toward thought-provoking and critical inquiry-based biological education. Now more than ever, in a political climate that asks us not to pull our students' lived experiences into the classroom (and sometimes threatens to punish us for it), we must consider how we as educators can act in disruptive and subversive ways to inspire criticality in our STEM classrooms.

My classroom experiences as both student and teacher, as well as my dissertation work as a Ed.D. candidate, have driven me to understand critical and queer pedagogy as key center points of my pedagogical practice. Critical queer pedagogies ask us as teachers to disrupt normative notions of the purpose of education and normative narratives about whose voice belongs in the classroom and when. These are critically important actions in creating subversive and disruptive STEM spaces for students, and they form the throughline of my reflections on disruptive STEM education.

In this chapter, I begin with an overview of critical pedagogy and queer theory. I then place these theoretical positions in conversation with how I am currently grappling with my positioning as a disruptive STEM teacher to demonstrate how these theories inform my pedagogical style. I show how they help me infuse key factors of disruptive STEM teaching into my classroom, including interdisciplinary perspectives, the inclusion of the self, and criticality.

Queer and critical pedagogies in the classroom

Queer theory draws from an arena of scholars and was initially coined by Theresa De Lauretis in 1991 to capture how scholarship should look at and examine queerness. De Lauretis positioned queer theory as having three central tenets: first, that it is necessary to reject heterosexuality as the dominant societal norm around which all understanding of sexuality and gender is framed; second, that gay and lesbian studies and queer populations cannot be looked at as a homogenous monolith of experiences; and third, that in order to understand queerness, we must look at intersections between sexual and gender identity, race, ethnicity, class and a variety of other social categories (De Lauretis, 1991). It is notable that queer theory is not seen as a distinct academic discipline – rather it is characterized by its ability to connect to multiple academic fields (Watson, 2005, p. 68), such as sociology, education, race and class history, cultural theory, feminist theory, literature, and more (De Lauretis, 1991, p. xvi). Queer theory seeks to position queerness as a point of view and a politic that challenges preconceived notions of "normal" within a societal context (Shlasko, 2006).

Since the inception of queer theory in 1991, many scholars have applied the tenets of queer theory to the sphere of education and pedagogy. When applied to educational endeavors, queer pedagogy asks us to challenge the ways that pedagogical practices uphold norms and normative views in educational spaces (Shlasko, 2006). This requires a reframing of educational content away from the traditional White, male perspective to call in identities of diverse races, genders, and sexualities (Pinar, 2003, p. 358). Second, queer theory asks educators to challenge away from an emphasis on *what* we teach and toward thinking about *how* learning is produced in the classroom in the first place.

The idea of disrupting normative pedagogical practices existed well before the conception of queer pedagogy. Queer pedagogy exists as a branch of the broader world of critical pedagogy. One of the most notable critical pedagogues is Paulo Freire. In *Pedagogy of the Oppressed*, Freire argues for liberation education, which rejects the notion of students being "banks" into which teachers deposit knowledge. Instead, Freire's pedagogical outlook posits that students are both teachers and learners, allowing students to bring in their own perspectives, experiences, and expertise and to partner with their teachers as equals in the learning experience (Freire, 1970, p. 72–73). Maintaining the "banking" system of education allows historically privileged voices to dominate the content taught in classrooms and allows educators to "deposit" normative ways of thinking into students. Challenging this structure of education allows for the transformation of the entire educational structure, so that oppressed voices are not simply integrated into the classroom but rather that the classroom is restructured to center oppressed voices, creating opportunities for critical thinking about the world at large (Freire, 1970).

Scholar bell hooks builds on Freire's work by engaging with freedom and education through a Black feminist and intersectional lens (hooks, 1994, p. 6). In her writings, she discusses how Freire's concepts of conscientization impacted her understanding of the self and identity as crucial to unpacking education as a tool for liberation. Still, she also remains aware of and directly critiques the sexism threaded through the language Freire uses in his writing (hooks, 1994, pp. 146–147). In doing so, she leans on criticality to both build her critique of Freire's work and finding value in his discussions on critical pedagogy. In *Teaching to Transgress* (1994), hooks argues for a pedagogy that is "anticolonial, critical and feminist" and that expands beyond traditional educational boundaries to interrogate supremacy and biases in classroom curriculum and to generate actionable pedagogical tools for teaching diverse students (p. 10). In line with queer theory, hooks also calls on educators to examine the intersections between identities and how they affect our engagement in disrupting normative narratives. In her writings, hooks calls for reframing learning as a "practice of freedom" and discusses the classroom as a space of "radical possibility" that can be transformed to call in intersecting identities and create spaces for liberation (p. 12).

In each of these critical pedagogies, the goal is clear: to disrupt normative pedagogical practices in order to create a liberated educational sphere that intentionally calls in historically marginalized voices and identities. It is through this lens of queer theory and critical pedagogy that I position myself as a disruptive STEM educator. Pedagogues can use queer theory and critical pedagogy as foundational theoretical tools in the classroom to ground the challenging normative views and standards in classrooms and to seek out new ways of being and doing in the classroom. As a critical pedagogy, queer pedagogy presents the idea that we can rethink what schools ought to be for. Queering the positionality and purpose of schooling can lead to teaching practices that challenge the norm and actively engage criticality, inclusion of the self, and interdisciplinary learning in the science classroom.

Inclusion of self

In studying the sciences, I was often taught that it is crucial to remove ourselves from the process of data collection and analysis, in order to be sure that the data is as free from error and influence as possible. I was told not to use "I" statements in writing up research methodologies and results, and it's rare that a social or human lens is pulled into the research. When I became a science teacher, this foundation stuck with me in major ways. Through my first few years, I found myself grappling with a tension of wanting to make biology as relevant and interesting to my students' lived experiences as possible while struggling to overcome the idea that the personal does not belong in the sciences. This tension required me to chip away little by little at my belief that personal experiences do not belong in the sciences. I found myself reflecting on and calling back to my own experiences as a high school student to engage in this unlearning process. While studying the sciences had pushed me toward removing the self from the science context, my first impactful biology learning experiences occurred when I was explicitly encouraged to include myself, my identity, and my ideas into my learning. My high-school biology teacher was the first person to open the door to the idea that as students, we mattered explicitly in the context of the science we were engaging in. Through reflections on these experiences, I began to discover how to honor the inclusion of myself and my students in my teaching practices.

I grappled with this tension significantly at the start of my educational career (and still do!). I keep the reflections on my own learning experience in the forefront of my mind, and it guides me in making decisions about inclusion of the self. I am a queer teacher, and I choose to be visibly queer within my science classroom. I incorporate discussions of gender, sexuality, and race and their connections to biological principles into my classroom each year, in a way that I hope allows students to see themselves as scientists and how science can be used to reject normative biological notions of race and gender.

In my experience, this is something students respond favorably to. Students thrive when they see themselves as belonging in the science classroom. Science is a subject that has often been understood as a White, male arena, and as disruptive STEM educators, it is our job to ensure that students understand the contributions of people of various backgrounds to the STEM field. Kim, Sinatra, and Seyranian (2018) found that educational interventions can change students' perception of what identities are a part of the science "in group" in the classroom. Through interdisciplinary curriculum construction and engaging in subversive teaching actions, inclusion of the self can be achieved in real and authentic ways for STEM educators and students alike.

In addition to the inclusion of my own identities, it is imperative to find ways to consistently and authentically incorporate the identities of the students into the science classroom as well. Students will always carry their lived experiences into the classroom, and as their teachers, they will oftentimes look to us to help them unpack and understand the world that is unfolding around them.

I started teaching in the fall of 2016, with the 2016 presidential election as a backdrop. My students were glued to the news, anxious to know what the result of the election would be. When we returned to school the day after the election, the fear created in many of my students was palpable. I was teaching children of undocumented individuals, who were fearful that their parents would be targeted to an even greater extent during the Trump presidency. As we moved forward into the school year, the consequences of the election began to unfold. In January 2017, a travel ban was implemented that targeted countries with large Muslim populations (Equal Justice Initiative, 2018). Students in my classroom feared this ban. Many were afraid that they as Muslim people in the United States would be targeted next. Others were fearful that they would never again be able to see their family members who lived in the targeted countries. Time and time again, students brought their lived experiences into my classroom as they looked for ways to understand what was happening to them and to the world around them.

I understood quickly that my job wasn't simply to teach science. My job was to help my students understand the world around them, to give them tools to ask questions, and push back against the injustices they were watching unfold in their own lives and the lives of their peers. My job was to create a space where students felt safe enough to express the fears and ask questions and grapple with them in meaningful and critical ways. This trend has persisted through my entire teaching career. During the COVID-19 pandemic, the Black Lives Matters movement, legislative infringements on the rights of LGBTQIA+ people, every teacher has had students sitting in front of them looking for answers to their questions, looking for support amid their fears, looking for ideas about how to move forward. As pedagogues and disruptive STEM teachers, interweaving themes of identity and the self, criticality, and interdisciplinary work can help us create spaces where we can help our students tackle their pressing real-world questions within the context of our discipline. Honoring our students' lived experiences and how they are impacted by the systems around them is an imperative function of our job as disruptive STEM educators. Examples of interweaving identity into the science can be found in subsequent sections of this chapter.

Subversive teaching in the biology classroom

In *Teaching as a Subversive Activity*, Postman and Weingartner contend with the concept that "change – constant, accelerating, ubiquitous – is the most striking characteristic of the world we live in and... our educational system has not yet recognized this fact" (1969, p. 4). They outline how governmental bodies and the media work to uphold normative stories and maintain status quo values. They put forward the idea that teaching, as a subversive act, can function as an "anti-bureaucracy bureaucracy" in which teachers teach students to look at their own society and ask *"why?"* and *"what is it good for?"* Postman and Weingartner note that these questions rarely find themselves at the surface in the media, politics, or research, and as a large group of folks who have access to students, teachers have a subversive responsibility to raise criticality in youth.

Subversively challenging dominant narratives and centering the question "why" fits perfectly into the world of science and queer pedagogy. For example, Broadway (2011) argues for a reimagining of biology instruction as an inherently queer pedagogy that challenges normative understandings of the biological world (p. 295). In this reimagining, science educators would support learners in deconstructing and reconstructing their understanding of their biological self and others (p. 300). In other words, queer pedagogy can be positioned in the science classroom in a way that pushes students to examine key scientific concepts and ask questions like "how do we know?" and "why do we view this concept in this way?"

Subversive teaching asks us to consider how we challenge norms and standards in schools while still operating within the system itself. One way this occurs in my classroom is in the space of teaching reproduction and heredity. The study of genetics is full of opportunities to subvert dominant narratives about gender and sex in a way that is directly standards-aligned, rigorous, and exploratory. One example is in discussions of sex chromosomes during units on heredity and genetic inheritance. This is a common topic found in nearly every high-school biology curriculum. In my biology classes, I have sought out a variety of ways to tease out the difference between sex and gender and to challenge the notion that there are only two biological sexes. I often start this sort of unit with explorations of the reproduction habits of species other than humans. Indeed, I am writing this chapter on the same day that I have just taught my students about reproduction in zebra sharks – an incredible species that can reproduce both sexually and asexually through a process called "parthenogenesis." We also discuss species of bacteria that reproduce by simply replicating their DNA and dividing to create two new offspring. Starting with discussions of species other than humans forces, the students and I work together to detangle our language of sex and gender. We talk about how discussing reproduction in gendered terms doesn't make sense for many species, whose reproductive habits are so different from our own.

This opens up the door to disrupt the vocabulary typically used in class to discuss human reproduction and heritability. We discuss the fact that not all individuals with uterine anatomy use the words "woman" or "mom" to refer to themselves, and not all people with testicular anatomy use the words "man" or "dad" to refer to themselves. From this positionality, we review how we will use scientifically accurate language and avoid using gendered language to discuss sex and reproduction in humans. This is a clear and subtle way to disrupt normative notions of gender and sex in the classroom while maintaining alignment with standards and helping students build the skills needed in the science classroom.

Interdisciplinary biology

A critical component in connecting queer theory and critical pedagogy to the science classroom is unraveling the sciences so that they are not seen as a standalone endeavor, but rather as deeply connected to other disciplines (Broadway, 2011). In the biological sciences, opportunities for contextualization and meaning-making of content through the lens of other subjects like history are always present. Biology, at its core, is the study of us. Biology allows us to understand how the world around us functions, from how organisms unknowingly collaborate to sustain trophic levels in ecosystems, to how the most microscopic cells and proteins in our bodies work together to convert raw materials into usable energy for our bodies. When examining current biology, it is necessary to see how biology has become an inherently interdisciplinary endeavor, incorporating key skills from a variety of disciplines.

However, it is challenging to find clear examples of how sciences can be taught in an interdisciplinary way at the secondary level (Nagle, 2013). One step in queering the teaching of biology is to create opportunities for students to ask critical questions about how they are biologically connected to the world around them. Broadway (2011) argues that positioning biology to "listen to other fields" is crucial to reimagining biology through a queer pedagogical lens and to identify other ways of knowing and questioning that can be pulled into the science classroom to create a transformative science learning experience (p. 300).

One way to put biology in conversation with other disciplines is through intentional cross curricular interdisciplinary projects. Contextualizing biology in other disciplines, such as history or literature, can help students examine the ways in which science is an endeavor that affects our experiences of ourselves and the people around us. This contextualization can be helpful in pushing students to ask critical questions about instances when the sciences have contributed to liberation, progress, and oppression, sometimes simultaneously.

When teaching in New York City, I was lucky to work with a grade team that was committed to interdisciplinary endeavors. As teachers in an Expeditionary Learning (EL) education school, one of the major expectations of our teaching craft was that we would engage in real-world and interdisciplinary problem-solving with our students. As such, we were constantly looking for ways to help students understand how different classroom subjects intersect with the real-world and with each other. As a team, we sought out connections between our content standards that would help us construct curriculum and standards-aligned interdisciplinary projects to cap each semester. The key examples we found were often connected to human rights. We selected topics such as the Columbian Exchange and the UN Declaration of Human Rights. These topics allowed us to put biology and history in conversation with each other while also leaning on skills from English language arts (ELA), math, the arts, and other sciences to complete their final projects. Rather than solely focusing on the technical content of biology, putting biology in conversation with history and literature allows students to use their scientific knowledge to ask questions about science and its relationship to human history.

A great example of this intersectional lens is the study of the UN Declaration of Human Rights and the history of scientific racism and homophobia. The UN Declaration of Human Rights includes a provision that entitles all humans to the right to life and the right to health. The UN also defines the right to benefit from scientific progress as a human right (United Nations, 1948). While rights to life and health, consent, and scientific progress are now enshrined in the UN Declaration of Human Rights, scientific investigation has historically been conducted in ways that have harmed marginalized people through scientific racism and sexism.

Examples of scientific misconduct abound. Louis Agassiz, a founding member of the National Academy of Sciences (NAS), prominently promoted false, scientific theories that used biological anatomy to claim that African people were an "inferior and separate race," a line of thinking that was used to justify the enslavement of Black people (Graves, Kearney, Barabino, & Malcom, 2022). Inaccurate scientific data were also used in the 19th century to argue against interracial marriage (Bergeron, 2021). A 2015 study showed that as many as half of medical students and residents problematically believe that Black women have naturally higher levels of pain tolerance than White women, leading to a major discrepancy in treatment quality (Hoffman, Trawalter, Axt, & Oliver, 2016). Homosexuality was listed as a mental illness in the Diagnostic and Statistical Manual (*DSM*) used by psychologists until 1973 (Drescher, 2015), while "transsexualism" was included as a mental health disorder from 1980 to 1994, when it was renamed as "gender disorder in adolescents and adults." In the most recent edition of the *DSM*, non-normative experiences of gender identity are referred to as gender dysphoria, and trans-individuals continue to need approval from a psychologist to undergo gender-affirming health care (American Psychiatric Association, 2013; American Psychiatric Association, 2023). This intersection between history and science provides a crucial opportunity for students to understand the interdisciplinary nature of biology and how it has been used as a tool of progress as well as a tool of oppression.

My colleagues and I used this lens in our collaborative interdisciplinary project in NYC, allowing students to place their scientific knowledge in the context of human rights to turn it over and examine it in a variety of ways. This project was completed with 10th grade students enrolled in AP Biology, AP World History, and Sophomore English. Students began this project by reviewing the UN's Universal Declaration of Human Rights, a document they had already encountered in their history class earlier in the year. We unpacked the document and identified possible ways that it could connect to biology and the sciences. Students then paired up with a partner based on the topics they were interested in studying. This created an opportunity for students to research a human rights science topic they were passionate about, including the right to informed consent, the right to medical care, or the right to a healthy environment. Students conducted research on a variety of topics, including recent bans on gender-affirming health care, the Tuskegee Syphilis cases, and medical experimentation and nonconsensual sample acquisition in cases like Henrietta Lacks and in the Uighur re-education camp crisis in China. To assist students who may have struggled to identify a possible topic, a list of ideas was provided to support them in choosing a topic they felt most interested in (Table 6.1).

In addition to offering differentiated options and substantial student choice, this project cracked-open science research to be looked at in a different way. Students were asked to select a topic that was meaningful or interesting to them, placing them directly at the center of their own research. Furthermore, a project of this type allows students to see the deep relationships between science and other fields of study. By leaning into interdisciplinary education,

TABLE 6.1 Human rights in the sciences curricular example

Learning Target: I can analyze and propose solutions to human rights questions in the sciences

Task: In this project, you will select an example of a human right, and research the ways in which it comes up in science. After doing research to identify key examples of how this human right interacts with scientific study, you will write a proposal that explains why this human rights issue is important to consider and how you think the scientific community ought to address this human rights issue. You can address your proposal to:

- A specific scientific community (the CDC, WHO, UN Science Panel, etc.)
- A governmental or congressional body that deals with human rights
- A student audience who is learning about human rights

You and your team members will create a Google site that communicates your research and proposal.

Possible human rights issues and examples to study (you are NOT limited to this list!)

Right to health and health care/right to bodily autonomy in medicine-Tuskegee syphilis experiments-Abortion rights-Universal health care	**Right to environmental protection**-Indigenous land rights-Flint water crisis-Air quality	**Right to food**-Genetically Modified Food Organisms-Equitable distribution and growth of food resources
Rights and protections of vulnerable populations-Discrimination in sciences on the basis of race, gender, and/or sexually	**Right to benefit from scientific progress**-Gene editing technology-Biotechnology-Gene therapy to treat diseases	

we taught students how to use learned concepts to solve problems as they are situated in the particular contexts in which they occur (Nikitina, 2007). By creating connections between different fields, we were "queering" the strict division of content and disrupting the idea that each content area ought to be siloed in its own way of learning and thinking. This approach creates rich investigations that center student voices, thereby inviting students' lived experiences and identities into the science classroom.

In my classroom, projects like these excite my students. They often choose topics and ideas that align closely with their own identities and experiences. Providing student choice that allows for the exploration of the self in an interdisciplinary way is one of the core ways that this project engaged students. They started to see how their different courses connect in very specific and relevant ways to their own identities. When engaged in real-world, interdisciplinary, and identity-connected science, students are involved, active, and produce high-quality, thoughtful, and rigorous final products.

Conclusions and final reflections

As bell hooks wrote, critical pedagogical practice allowed her to "…imagine and enact pedagogical practices that engage directly both the concern for interrogating biases in curricula that reinscribe systems of domination (such as racism and sexism) while simultaneously providing new ways to teach diverse groups of students" (1994, p. 10). I argue that the application of queer theory and critical pedagogy can help provide a theory-driven approach to disruptive STEM teaching in the classroom that similarly interrogates biases within the sciences while creating significant opportunities to build skills in diverse student groups.

The current political backdrop in the United States makes this work even more imperative and even more challenging. The growing number of bills that ban books and prevent discussion of race, gender, and sexuality in the classroom create conditions that can make it dangerous for teachers to engage in critical discussions of identity and diversity with students. In turn, this can create unsafe conditions in the classroom for students who have questions about themselves and their histories. In a time when discussion of identity and self are being explicitly banned in the classroom, subversive STEM teaching becomes even more crucial. Subversive STEM allows us as pedagogues to align interdisciplinary critical questions about the world to rigorous skills and learning standards.

Subversive and disruptive STEM teaching asks us to find ways to persist in the face of challenges to the most important aspects of our classrooms. As a subversive STEM pedagogue, my goal is to help my students see the value in science as a space to investigate themselves and the world around them in a meaningful way. I hope that by providing a space where students engage in interdisciplinary work, rigorous skill-building, and critical thinking about themselves and others, I create the conditions for students to question the things they learn – both while they are in and after they leave my classroom.

References

2017 travel ban. (2018, March 29). *Equal justice initiative.* https://eji.org/news/history-racial-injustice-2017-travel-ban/

ACLU. (2023, May 2). *Mapping attacks on LGBTQ rights in U.S. state legislatures.* American Civil Liberties Union. Retrieved May 2, 2023, from https://www.aclu.org/legislative-attacks-on-lgbtq-rights

Alexander, T., Baldwin Clark, L. T., Reinhard, K., & Zatz, N. (2023). (rep.). *CRT forward: Tracking the attack on critical race theory.* UCLA. Retrieved from https://crtforward.law.ucla.edu/wp-content/uploads/2023/04/UCLA-Law_CRT-Report_Final.pdf

American Psychiatric Association. (2013). *Diagnostic and statistical manual of mental disorders* (5th ed.). American Psychiatric Publishing.

American Psychiatric Association. (2023). *Gender dysphoria diagnosis. Psychiatry. org - Gender dysphoria diagnosis.* Retrieved May 2, 2023, from https://www.psychiatry.org/psychiatrists/diversity/education/transgender-and-gender-nonconforming-patients/gender-dysphoria-diagnosis

Bergeron, E. (2021, July). *The historical roots of mistrust in science.* American Bar Association. Retrieved May 2, 2023, from https://www.americanbar.org/groups/crsj/publications/human_rights_magazine_home/the-truth-about-science/the-historical-roots-of-mistrust-in-science/

Broadway, F. S. (2011). Queer (v.) queer (v.): biology as curriculum, pedagogy, and being albeit queer (v.). *Cultural Studies of Science Education, 6,* 293–304. https://link-springer-com.ezproxy1.lib.asu.edu/content/pdf/10.1007/s11422-011-9325-7.pdf

De Lauretis, T. (1991). Queer theory: Lesbian and gay sexualities. *A Journal of Feminist Cultural Studies, 3*(3), iii–xviii. 1040–7391.

Drescher, J. (2015). Out of DSM: Depathologizing homosexuality. *Behavioral Sciences, 5*(4), 565–575. https://doi.org/10.3390/bs5040565

Freire, P. (1970). *Pedagogy of the oppressed.* Penguin Education.

Graves, J. L., Kearney, M., Barabino, G., & Malcom, S. (2022). Inequality in science and the case for a new agenda. *Proceedings of the National Academy of Sciences, 119*(10). https://doi.org/10.1073/pnas.2117831119

Hoffman, K. M., Trawalter, S., Axt, J. R., & Oliver, M. N. (2016). Racial bias in pain assessment and treatment recommendations, and false beliefs about biological differences between blacks and whites. *Proceedings of the National Academy of Sciences, 113*(16), 4296–4301. https://doi.org/10.1073/pnas.1516047113

hooks, bell (1994). *Teaching to transgress.* Routledge.

Kim, A. Y., Sinatra, G. M., & Seyranian, V. (2018). Developing a stem identity among young women: A social identity perspective. *Review of Educational Research, 88*(4), 589–625. https://doi.org/10.3102/0034654318779957

Nagle, B. (2013). Preparing high school students for the interdisciplinary nature of modern biology *CBE—Life Sciences Education, 12*(2), 144–147. https://doi.org/10.1187/cbe.13-03-0047

Nikitina, S. (2007). Three strategies for interdisciplinary teaching: Contextualizing, conceptualizing, and problem-centring. *Journal of Curriculum Studies, 38*(3), 251–271. https://doi.org/10.1080/00220270500422632

Pinar, W. F. (2003). Queer theory in education. *Journal of Homosexuality, 45*(2–4), 357–360. https://doi.org/10.1080/00220270500422632

Postman, N., & Weingartner, C. (1969). Teaching as a subversive activity. *Popular educational classics.* https://doi.org/10.3726/978-1-4539-1735-0/17

Shlasko, J. D. (2006). Queer (v.) pedagogy. *Equity & Excellence in Education, 38*(2), 123–134. https://doi.org/10.1080/10665680590935098

United Nations. (1948). *Universal declaration of human rights.* The United Nations General Assembly.

Watson, K. (2005). Queer theory. *The Group Analytic Society (London), 38*(1), 67–81. Sage. 10.1177/0533316405049369

7

THE CITY BUDGET PROJECT

Proportionality, financial literacy, and culturally responsive mathematics

Enrique Nuñez

Introduction

What do you think of when you hear the word *math*? Do you think of equations, division, or bar graphs? What about the Pythagorean theorem? Do you think of Pi Day? Regardless of how you answer, I assume you think of math as a concept that follows procedural computational skills and structured rules that build an understanding of mathematical concepts needed for problem-solving. Take the Pythagorean theorem. The *Encyclopedia Britannica (2023)* defines it as a theorem in which the sum of the lengths of the legs of a right triangle squared equals the square of the hypotenuse, or, put in algebraic notation: $a^2 + b^2 = c^2$. This has been the general rule since its assumed founding by Greek mathematician-philosopher Pythagoras, who lived from 570–500/490 BCE. I too used to believe that math was a black-and-white subject where there was a clear right and wrong answer that can be arrived at only if a person has computational skills and understands the structured rules and procedures of a given problem. Based on this framing of mathematics as a structured subject that lives within a world of constraints, procedures, and computations, is it possible that mathematics can be culturally responsive to the needs of learners with diverse backgrounds?

As a math teacher myself, I recognize that there is bias in the following statement: math is *the* most important subject taught. Regardless of personal bias, society has prioritized math and science knowledge because the world is becoming more technologically advanced. Additionally, pursuing and obtaining a math or science-oriented education has been described as having a larger financial payoff (Louka, 1993, Rocheleau, 1995). Performance in

DOI: 10.4324/9781003395782-11

K–12 mathematics has been framed as the ultimate gatekeeper for learners hoping to get into higher education and the educated workforce (Douglas & Attewell, 2017; Hunter, Hunter, & Bills, 2020). Simply put, math is a determinant in a learner's life development and progression (Hunter et al., 2020); yet, math education is failing learners of diverse backgrounds in urban school districts across the country. This failure produces learners who are then unable to be fully productive and contributing members of society and reach greater economic prosperity (Ukpokodu, 2011), stunting their life development and progression. The mathematics learning crisis for diverse learners from urban and low-income communities is "caused by school policies, curricula, and teaching practices that do not engage those students" (2011, p. 48). Only when math curriculum, instruction, and assessments are centered on the lived experiences, cultures, and communities of learners with diverse backgrounds will those learners be able to become engaged and empowered by their mathematical experience in the classroom (Asante, 1991; Tate, 1995; Ukpokodu, 2011).

In this chapter, I utilize theoretical approaches that empower learners of diverse backgrounds in mathematics education. Focusing predominantly on Gay's (2000) coining of culturally responsive pedagogy and Ukpokodu's (2011) experience working with teachers' inquiries regarding culturally responsive teaching, I articulate the ways in which I have attempted to be culturally responsive in my current position as a 7th-grade mathematics teacher.

This will be my 4th year as a mathematics teacher in the city of San Antonio, Texas; my role has been of great importance to me because I work in the community that I was born and raised in; while I did not attend the school I teach at, I attended schools within the school district I now work for. My district is an urban public school district encompassing a large part of San Antonio's inner city, where a large percentage of its population consists of learners from low-income Latinx households – myself included. Unlike the majority of schools within my district, the school I work for, that I am giving the following pseudonym, The Emma Tenayuca School for Young Women, is an all-girls public school, the first in San Antonio whose mission statement is to support young women who seek a college-readiness public education that will equip them with the academic and social skills to achieve success in college and life through critical thinking, leadership, and social-emotional wellness. As a queer man of color, I have an immense responsibility to empower the predominantly young women of color I serve by creating a space where they are not only able to succeed academically but develop a love for learning and can apply their knowledge to their own lives. The Emma Tenayuca School focuses on math, science, and technology because, historically, the science, technology, engineering, art, and math

(STEAM) fields have been underrepresented by women. I have created a classroom environment that empowers the young women in my classroom to become passionate about STEAM.

In addition to academics, The Emma Tenayuca School aims to develop responsible leaders within our learners, leaders who are active members of their communities. In committing to the curricular focus and mission statement, I created the "City Budget Project." Through the creation and implementation of an activity, the "City Budget Project," draws on Texas Education Knowledge Skills (TEKS) from units related to proportionality and financial literacy, I have found ways to address TEKS state standards and learning objectives that learners *must* know while simultaneously creating an activity responsive to my learners' communities and to events related to our current sociopolitical climate. In this chapter, I draw on the work of Geneva Gay (2002), who argues that culturally responsive teaching can influence the academic achievement of diverse learners, the creation of the "City Budget Project," draws on learners' knowledge of mathematics content, but, most importantly, positions students to be critical of the ways in which the city of San Antonio distributes funds within its budget across different city institutions.

The City Budget Project

In alignment with the Emma Tenayuca School's vision to promote critical thinking, leadership, and social skills, and my desire to bring social justice to mathematics, I created the "City Budget Project." The goal of this project is to have learners identify and evaluate the proportional relationship or lack of it, within the San Antonio city budget over the past three fiscal years. Learners had to focus on the increase or decrease of funding of three city-funded public institutions such as parks, libraries, police and fire, and health. Learners used mathematical concepts and conceptual applications related to proportions, rates of change, and budgets to visually display their data and findings on a graph.

Once learners have gathered their data, they are tasked with writing a letter. This letter includes why they chose their respective institutions, their findings, the visual graph they created, and an argument as to how they would reallocate money from our city budget. This letter was then addressed to city officials such as the mayor or learners' respective councilperson. In short, the "City Budget Project" asks learners to answer the following question: Is funding for public institutions within the city of San Antonio increasing or decreasing at a proportional rate? At the heart of this project is the culminating letter writing. Letters were rooted in my learners' lived experiences and detailed how public institutions impact their lives. Here is where the "City Budget Project" enshrines the importance that cultural pedagogies for diverse learners can play in the classroom, especially in a math classroom.

Cultural pedagogies for learners of diverse backgrounds

Gloria Ladson-Billings (1995) defines culturally relevant pedagogy as a "theoretical model that not only addresses learner achievement but also helps learners to accept and affirm their cultural identity while developing critical perspectives that challenge inequities that schools (and other institutions) perpetuate" (1995, p. 469). Culturally relevant teachers believe all learners, regardless of their sociopolitical identity, can show academic achievement; acknowledge the diverse identities of learners; demonstrate cultural competence; empower learning through the usage of cultural referents; and create an environment where both teachers and learners can critique societal injustice (Ladson-Billings, 1995; Mensah, 2021).

In my math classroom, I draw on the work of Geneva Gay by including nuanced thinking, the creation of the "City Budget Project," aims to have learners learn in alignment with the TEKS learning objectives while also asking students to think of the nuanced ways in which city funding prioritizes certain institutions over others such as police and fire over city infrastructure. Geneva Gay furthers Ladson-Billing's work through culturally responsive teaching, defined as "using the cultural characteristics, experiences, and perspectives of ethnically diverse learners as conduits for teaching them more effectively" (Gay, 2002, p. 106). Culturally responsive classrooms are environments where learners' lived experiences and contexts are valued and seen as pedagogy, thereby creating a classroom space that is engaging, meaningful, and personable (Gay, 2000). Culturally responsive teaching asks teachers to step outside of an objective and neutral mindset because our world is nuanced and requires critical thinkers to step away from traditionalist thinking; this is especially true in the context of mathematics – a subject that is framed as right or wrong (Ukpokodu, 2011). Much like culturally relevant classrooms, a culturally responsive classroom's pedagogical usage of learners' cultures and lived experiences has proven to increase learners' academic achievement (Gay, 2000, 2002; Ladson-Billings, 1995). Gay's work supports the goals of my City Budget Project because learners must use their cultural experiences as the driving force for their learning. Whether they intended to do so or not, learners were not neutral when picking their three institutions, the majority chose institutions that directly impacted their lives. They chose health, parks, libraries, police and fire, transportation, and so on.

Due to the diverse backgrounds of my learners as mostly women of color in addition to my own cultural identities, I located this work within a broader framework of culturally sustaining pedagogy. Derived from Ladson-Billings and Gay's work, Django Paris (2012) coined the concept of culturally sustaining pedagogy. Culturally sustaining pedagogy (CSP) "seeks to perpetuate and foster – to sustain – linguistic, literate, and cultural pluralism as part of school for positive social transformation" (Paris & Alim, 2017, p. 1). CSP is an extension of Ladson-Billings and Gay's theoretical approaches, arguing that schools and teachers must not only affirm and connect with learners' cultural backgrounds but sustain them within the classroom (Paris, 2012; Paris & Alim, 2014). CSP works against the eradication of learners' cultural ways of being, therefore disrupting the ways in which schools operate under the norms of Whiteness and colonialism (Paris & Alim, 2017). Learners in most public schools in the United States are provided with a simplified curriculum wherein they cannot see themselves or feel as though their experiences are incorporated into their daily learning. Thereby aggregate school environments are created where learners of diverse cultural backgrounds are seen as lacking in academic ability (Mensah, 2021). While each of these three theoretical approaches is unique, together they push back on theoretical frameworks that paint math as a finite subject that only exists within the realm of higher education and the educated workforce (Douglas & Attewell, 2017). Instead, Ladson-Billing's, Gay's, and Paris' theoretical approaches offer a shift in the social construction of mathematical education toward being a space that empowers learners of diverse backgrounds. When implemented within the classroom, culturally relevant pedagogy, culturally responsive teaching, and culturally sustaining teaching have been proven to increase learner academic achievement and engagement in school. I've noticed in my own practice that my classroom becomes a site of identity affirmation where I can build a community of learning by encouraging learners to use their lived experiences as the driving force of their learning. While the "City Budget Project" is in alignment with learners' mathematical course knowledge, the project draws on the cultural pedagogies mentioned above because it is driven entirely by the experiences of the learners in my classroom. Learners' lived experiences are sites of knowledge, without them, they would not be able to truthfully complete the "City Budget Project."

Embedded in subjects such as English or history, these pedagogies create space for learners to see themselves and their communities reflected in texts read for English or through first-person historical accounts provided by people of marginalized communities for history classes. While it does take work, I argue that embedding culturally relevant, culturally responsive, and/or culturally sustaining pedagogies within English and history curricula is easier than embedding them into mathematics (Ukpokodu, 2011). Simply put, there is more content that can be gathered within English and history classes created by people of diverse cultural backgrounds that align with the principles of the theories laid out by Ladson-Billings, Gay, and Paris. English and history can and have been studied and interpreted in a multitude of ways, allowing learners to theorize and reflect on the beliefs, ideas, histories, and stories of diverse people to explain how society operates thereby allowing learners to critique societal systems of inequity and create action plans for a better world. Unlike these subjects, the rules and procedures of mathematical concepts have little room for change.

Before continuing, it should be noted that the prior statements are not said with malicious intent, especially in today's political climate. Many Republican-led legislatures are introducing laws aimed at regulating how teachers in the United States discuss issues related to systemic inequality such as racism, sexism, and homophobia in the classroom (Schwartz, 2021). These courses are most vulnerable to "anti-CRT" and the "Stop W.O.K.E" bills passed in Florida that are now beginning to infect other parts of our nation. Math is relatively safeguarded from this political controversy because, to many, mathematics is simply a set of rules and procedures learners must follow in order to get the "right" answer; math is the gatekeeper and focal point of social mobility.

Can mathematics be culturally responsive?

If you ask most math teachers if mathematics is a subject in which culturally responsive pedagogy (CRP) can be embedded, the answer would be *no*. In Ukpokodu's (2011) exploration of teachers' inquiries regarding CRP in mathematics, teachers expressed why they were not engaged in culturally responsive mathematics teaching. Four major themes emerged:

1 View of mathematics as culturally neutral
2 Convenience and dominance of textbook-based mathematics instruction
3 Curriculum standardization and high-stakes testing
4 Lack of culturally responsive teaching models to emulate (pp. 49–50).

First, as expressed at the beginning of this chapter, mathematics is usually considered culturally neutral, or a "universal language" in the words of one of Ukpokodu's research participants (2011, p. 50). This "universal language" portrays math as a fixed concept that is the same across the world, with rules that have remained unchanged and unchallenged. With the exception of its origin, the Pythagorean theorem is a rule that has been left unchallenged since the lifetime of Greek philosopher Pythagoras who lived circa 500–400 BCE. Second, the convenience of textbooks within mathematics lightens teachers' workloads by providing them with a set curriculum to teach, often making it harder and/or more time-consuming to include CRP (2011). Third, mathematics teachers must teach with the goal of increasing learners' standardized test scores in mind, and therefore instruction pivots to teaching to the test. "Because of this pressure to raise test scores, many teachers feel restricted and powerless to teach in ways that are culturally responsive to meet their students" (2011, p. 50). These three themes, coupled with the sheer lack of culturally responsive mathematical teaching models to emulate, have made it difficult for teachers to embrace a mathematics classroom wherein CRP can be used to engage learners from diverse backgrounds and help them develop a passion for mathematics and its applications within the everyday lives.

To be clear, I do not have all the answers on how to become a culturally responsive math teacher, nor do I have an abundance of material or teaching models for others to emulate. I am a work in progress because, like all things, being a culturally responsive mathematics teacher is an ever-evolving feat – there is no finish line. Culturally responsive teaching must continuously evolve because of the lived experiences and identities of the learners that we teach and the sociopolitical events unfolding each day. To pay more attention to the lives of Black learners over other learners of color or to focus on a surface-level contextualization of all women over the experiences of women of color ignores the multifaceted and intersectional identities that exist within our classrooms; in doing this, we minimize the forms of knowledge and experiences of learners with by multifaceted identities (Gay, 2002). In short, being a culturally responsive educator is no easy feat, but we must start somewhere.

For me, the starting point was Gay's (2002) connection of culturally responsive teaching to building culturally responsive communities of learning where, as she states

> Personal, moral, social, political, cultural, and academic knowledge and skills are taught simultaneously... Culturally responsive teachers help learners to understand that knowledge has moral and political elements and consequences, which obligate them to take social action to promote freedom, equality, and justice for everyone.
>
> *(2002, p. 110)*

At the core of every culturally responsive learning environment should be a desire to work to uplift and honor the communities and experiences of our learners. The activity I created asked learners to analyze city budgets over the course of three years to determine if allocations for city departments were increasing or decreasing at a proportional rate. This activity serves the learning goals associated with state standards and skills related to proportional reasoning, financial literacy, and data analysis while also elevating mathematical literacy in that learners are able to communicate computational findings and from that, develop arguments in the form of a letter to a city official wherein they presented their findings as to whether funding was distributed at a (non) proportional rate and argue where they believe funding should be allocated. This culminating letter in turn makes the "City Budget Project" an activity connected to cultural pedagogies of knowledge, particularly to culturally responsive teaching. While situated within my pacing schedule's unit on financial literacy, the idea for the "City Budget Project" came to fruition in response to the dialogue surrounding how capital flows through communities, a dialogue that was only intensified in 2020 in response to the COVID-19 pandemic, and the murder of George Floyd, a Black man, by Minneapolis Police officer Derek Chauvin. This project allows learners to identify which institutions are prioritized by city leaders while simultaneously identifying which institutions are most important to them based on their lived experiences therefore arguing for redirection of the flow of capital in the city of San Antonio budget.

The City Budget Project and culturally responsive teaching practices

In addition to asking her participants *why* teachers were not engaged in culturally responsive mathematics teaching, Ukpokodu (2011) also asked participants *what* culturally responsive teaching practices are. Their responses focused on seven major themes, including:

1 Deconstruct misguided beliefs about mathematics teaching and learning
2 Integrate culturally relevant content and social and justice issues
3 Utilizing culturally responsive instructional strategies
4 Foster communal learning
5 Openness to students' divergent thinking and problem-solving
6 Detrack the mathematics classroom
7 Teacher's critical consciousness, advocacy, and activism (2011, p. 50).

I use these seven major themes to address what CRP has looked like for me in my 7th-grade math class. I teach at a public all-girls school that consists of learners from all across the greater San Antonio region. While my classroom accounts for a single-gender population, not all of my learners identify as women, and there are learners of different racial and ethnic identities and of different socioeconomic statuses. When given the chance to engage with one another as to which institutions were of most value to each of my learners, conversations were situated around the varying economic and geographical status of my learners; these identity markers were the driving force of learners' project submissions. San Antonio is a racially and therefore economically segregated city, because the Emma Tenayuca School consists of learners from across the city and the greater Bexar County area, learners' lived experiences vary. A learner who grew up in the "West Side," or 78207 zip code, a neighborhood that has historically consisted of low-income households, with some living below the poverty line, will not have the same lived experiences or prioritize the same institutions that a learner from "Olmos Park," one of the wealthiest neighborhoods in San Antonio.

The "City Budget Project" tasked learners with analyzing the city of San Antonio's budget to determine if the distribution of city funds to city institutions over the course of three years was proportional. Logistically, learners worked on the project over the course of two weeks within four ninety-minute class periods. Learners utilized proposed budget documents ranging from the 2018 to 2023 city of San Antonio fiscal years from the Office of Management and Budget. While learners submitted individual projects, they were encouraged to share their findings and opinions with one another; many shared an appreciation for being able to learn from another as well as a building of empathy in that many were provided more insight into the lives of their peers. Once learners have recorded their findings after analyzing city budgets for proportionality in budget allocation across three city institutions, they are tasked with writing a letter to city officials sharing their findings and detailing whether they believe funding is distributed fairly and arguing in support of why other institutions of their choice should have more funding allocated toward it. While the letter was not sent to the city officials due to time constraints and end-of-the-school-year procedures, the inclusion of a letter as the culminating piece of the project provided learners with the opportunity to make an argument rooted in evidence to improve the outcomes of their communities and our greater city of San Antonio. This letter is an act of political significance because learners are encouraged to engage in civic dialogue as to how they wish for capital to flow across city institutions. Using this project as an example in conjunction with the seven major themes, I hope to add to the culturally responsive teaching models that math teachers across the country can emulate within their own classrooms.

Deconstruct misguided beliefs about mathematics teaching and learning

The "City Budget Project" is the culminating project within our unit on financial literacy that comes at the end of the school year. Drawing from the disruptive STEM framework, specifically the subversive element, I positioned this lesson at the end of the academic year after students took their standardized tests allowing this activity to have fewer constraints. Since this unit is not heavily tested on standardized tests, it is a low-priority item. Because of this, I felt it was the most appropriate place to begin my journey as a culturally responsive mathematics educator. Financial literacy is one of the units I teach that seems to touch the lives of learners the most because of how quickly they can connect it to their lives. When introducing budgets, learners discuss how their parents budget for items like food, housing, and transportation. Some learners also discuss how debt plays a huge role in their parents' budget, while others discussed budgeting when parents have low-income jobs; or how parents may overspend in some categories compared to others. In this way, I use *cultural scaffolding* with my learners: through discussion of budgets and their applications to learners' lives, I use their own experiences to expand their intellectual horizons and recognize mathematics as having applications within their sociopolitical lives (Gay, 2002; Ukpokodu, 2011).

Conversations around budgets deconstruct misguided beliefs about mathematics being culturally neutral or universal (Ukpokodu, 2011): budgets are not universal. Our unit on financial literacy turns math into a subject that "attempts to describe and understand physical and social phenomena" (Mukhopadhyay & Greer, 2001). As mentioned earlier, the learners I teach have different socioeconomic statuses, and therefore conversations about budgets are different for each of them (Gay, 2002). Most learners of color state that their parents take into account large family gatherings such as *quinceañeras* within their budgets. Others shared that they do not see their mother often because, as a single mother, she must work long hours to make ends meet. One learner expressed discomfort discussing budgeting because she had never felt the strain of finances like other classmates; her family budgeted for lavish vacations in different countries. Each of these lived experiences offers insight into how and where mathematics can start to come to life.

Integrate culturally relevant content and social and justice issues

Participants in Ukpokodu's (2011) study identified ways to integrate multicultural or culturally relevant content, including "integrating social issues relevant to the learners' community" (p. 51). I created the "City Budget Project" during my first year as a math teacher during the academic school year of 2020–2021, a year that was embellished by two extraordinary circumstances playing out on the national and global stages. The ongoing COVID-19 pandemic and the *#BlackLives Matter* movement, created in

2013 by three Black women in response to the murder of 17-year-old Trayvon Martin, that organized protests against police brutality targeting Black people following the murder of George Floyd.

While the United States declared the COVID-19 Public Health Emergency over on May 11, 2023, the legacies of the COVID-19 pandemic will be felt forever, as this was an experience that most of the modern world had never faced before. The pandemic forced many to quarantine and stay home for months on end. At its onset, the US government and its health institutions could not meet the demands raised by the pandemic. Hospitals ran low on ICU beds and healthcare workers, highlighting the lack of medical infrastructure to meet the needs of COVID-19, other illnesses, or to prepare for the next pandemic.

School buildings closed, forcing learners to receive virtual instruction via platforms like Zoom. Districts like my own utilized district and federal funds to ensure learners had the resources to meet the needs of virtual instruction, as well as other needs, such as hot meals once provided within the physical school building. As federal funding for these COVID-19 response needs come to an end, school districts have faced a budget crisis due to the debt accrued in response to meeting learner needs. Since physically returning to the classroom, learners have brought with them an enlarged educational deficit, increased responsibilities for teachers, and a rise in mental health, concerns all of which have been met with little funding. The pandemic made many realize the sheer lack of readiness from the political, social, and *economic* spheres to respond to a large crisis. This would be further highlighted in the *Defund the Police* and *Black Lives Matter (BLM)* movements that arose in response to the murder of George Floyd.

Racism and inequitable treatment of BIPOC people, particularly Black lives, continues to plague our society, most notably through the continued violence against Black lives through the very organizations that are meant to keep people "safe:" *law enforcement.* The murder of George Floyd by Derek Chauvin, a police officer in the Minneapolis Police Department, in May 2020 sparked protests against police brutality, and furthered the call to "defund the police." While this call means different things to different people, at its core, it calls on society to reevaluate how policing in America has risen in the modern era predominantly through increases in funding for law enforcement (Henry & Wing, 2021; Vitale, 2017). Calls to "defund the police" include ideas like reallocating money from law enforcement to other city institutions and social services or investing in alternative public safety programs in an attempt to curb increased policing in cities across the country. While different in nature, both extraordinary circumstances call into question where our priorities land when investing in institutions to meet people's needs. Could we have been better able to meet the needs of people living in the United States throughout the COVID-19 pandemic? **Yes**.

Had we invested more money in schools, health care, and other institutions, we would have been better able to curb the effects of COVID-19 while decreasing the need for law enforcement, thus reducing police violence on Black bodies. The impacts of both COVID-19 and BLM have impacted everyone across all identities and ages, including school children. K–12 learners across the country too have lived through the pandemic; they too have seen via news outlets and social media or have participated in the movement against police brutality. In response, should they not be allowed an avenue to express their opinions on the matter? In positioning Ukpokodu's (2011) participants' claim that culturally responsive mathematics teaching must integrate social issues relevant to the learners' community, I created the "City Budget Project." This project serves as a response to national and global cries for federal governments to take political and economic action to address the needs created by the COVID-19 pandemic and reevaluate the drastically large funding of the police – I wanted to see how my own learners would reallocate money within our own city.

Utilizing culturally responsive instructional strategies

Culturally responsive mathematics teachers must be self-reflective and ready to critique and question their teaching praxis (Shor, 1992), in order to ensure that their classroom is inclusive and reflective of learner identities. Participants in Ukpokodu's (2011) study raised the following self-critiquing question: "Who is learning math in my classroom and who is not, and why?" (2011, p. 53). Learners of diverse backgrounds are failing mathematics because many people equate "smartness" and "worth" with performance on standardized tests in the subject. Learners who perform well on standardized tests thrive in my classroom because they are able to correctly answer questions on a test, shutting out those who do not perform on the same level. Low-performing learners are then tracked and grouped with other learners, who are then provided with lower-level work that does not encourage divergent thinking (Ukpokodu, 2011).

The "City Budget Project" utilizes culturally responsive instructional strategies because it deviates from tracking and grouping; the project includes *all* learners regardless of performance level. When introducing the project, it's framed as a conversation starter, asking learners:

- How do you define the word "budget"?
- How have you heard this word used within your household? What are things included within your household budget?

- Outside of the household, what other places use budgets?
- Focusing on city budgets, what institutions do you believe rely on city funding?
- Of these institutions, which is the most important to you and why?

This low-risk entry exercise is not off-putting because it does not ask learners to immediately delve into computational skills to answer a basic standardized test question. These questions ignite engagement and thinking for everyone regardless of performance level; all learners have at least one thing to share about budgets because of their application to learners' lives. They are able to speak openly and share perspectives on personal and city budgets, the city institutions that have value to them, and why.

Foster communal learning

In urban communities consisting of Black and Latino populations, many come from close-knit households composed of several generations within neighborhoods where many work and play together (2011). This is in alignment with many ethnic groups' cultural values as some, like Black and Latino communities, "give priority to communal living and cooperative problem solving [thereby affecting] educational motivation, aspiration, and task performance" (Gay, 2002, p. 107). Math must be a space where communal learning is embedded within the classroom culture; learners learn best when they are able to learn from each other, especially when working to grasp complex computational skills and procedures. Some of the best moments of learning with my learners have not been through guided or small group work under my supervision but when they are able to work together. Here learners are able to easily identify and address one another's misconceptions in user-friendly language that may take time for me, their teacher, to break down for them. Most importantly, this increases learner engagement.

While learners submit their own letters explaining their findings related to proportionality in budget allocations to city institutions, they are encouraged to work together to compare and contrast their opinions on which city institutions are of importance to them. They also discuss their choices of which institutions they used to determine if those institutions received a (un)proportionate amount of money over the course of four fiscal years, or whether they gained or lost relative funding. Together, learners are able to learn about the task at hand as well as the importance of different institutions within the lives of their peers. When actively monitoring the classroom, I overheard learners stating, "I never thought of Police & Fire that way" or "Wow! I didn't know that some of the money from Parks and Recreation went to protecting the Edwards Aquifer! My mom says that's where we get most of our drinking water from!" Through this activity, learners are able to gauge how the city budget impacts each other's lives: those who share the same cultural background as them and those who do not.

Openness to students' divergent thinking and problem-solving

For learners of diverse backgrounds, thinking and problem-solving can be divergent as many of these learners "contextualize their thinking and processing style based on their cultural and lived experiences" (Ukpokodu, 2011, p. 54). Responses were grounded in the cultural and lived experiences of the learner when responding to the questions:

- Focusing on city budgets, what institutions do you believe rely on city funding?
- Of these institutions, which is the most important to you and why?

One shared that the city should invest more money in schools. She used our school building as an example of needing funds as it is an older building that most recently had suffered water damage due to the increased spring rainshowers in the San Antonio area. Another shared that she believed the city should invest more money in housing and development because of the increased homeless community within her neighborhood. One even stated that there should be a larger investment in health care as people will always need to see the doctor and some cannot afford it due to a lack of health insurance. Open-ended questions like the ones posed above and in prior sections are important to include within the classroom because they decenter math as a "universal language." Prioritizing different categories and institutions within personal and city budgets is not universal, it looks different based on a learner's culture and lived experiences.

Detrack the mathematics classroom

While learners receive a grade on the "City Budget Project," their grade is not solely rooted in showcasing their abilities to use computational strategies to get the "right" answer. This serves to disrupt traditional practices. The project used a rubric that focused on the following criteria:

- Calculations represent (non) proportional rates of change in mathematical and real-world problems
- (Non) proportional relationships represented in a linear graph; Represent linear relationships using graphs

- Letter written to city officials follows Claim, Evidence, Reasoning (CER) formatting:
  - The letter states a claim
  - Effective mathematicians communicate their work within the evidence part of their letters written to city officials
  - The letter has a reasoning statement

Even though computational skills are a part of learners' overall assignment grade, they do not have to feel nervous that a lack of computational skills will be used against them. Computations (calculations) are one part of the grading rubric; they are graded holistically. Yet, due to the value of standardized test scores by the Texas Education Agency and my administration, I have to group learners based on instructional ability to target the needs of those who struggle the most. As such, the setup of this project provides me a reprieve to work with learners who struggle via instructional scaffolding (Ukpokodu, 2011). Regardless of instructional grouping, I do not water down the rigor of this project for learners at a "lower" track in this project – all learners are held to the same expectations. The rubric positions math as a subject that is not only based on "right" versus "wrong" but also creates a space for argument, drawing on other forms of knowledge that my "lower-level" learners already have: those derived from their cultural beliefs, lived experiences and frames of reference. Therefore, I do not lower expectations for them – all learners are exposed to the same mathematics curriculum (Gay, 2002; Ukpokodu, 2011).

Teacher's critical consciousness, advocacy, and activism

Culturally responsive mathematics teaching requires teachers to place critical consciousness, advocacy, and activism at the forefront of their practice (Ukpokodu, 2011). The US COVID-19 pandemic response and continued violence from law enforcement on Black bodies have continued to fuel my passion for a socially just society, one in which our government is able to respond to the needs of the people. I dream of a country in which people have access to medical treatment regardless of health conditions or income status. I dream of a country where law enforcement is held accountable for their actions and where policing comes to a decline. I agree with the call to *Defund the Police*; money and resources must be reallocated to social services to address the needs that led people to crime. Alternative safety programs for our communities that reject the police brutality against Black lives exist and must be funded (Henry & Wing, 2021; Vitale, 2017).

While time has passed since the onset of the COVID-19 pandemic and the murder of George Floyd, the lack of funding to address health needs and systemic inequities that fueled anger over those events still exist. This anger stems from the ways in which our federal, state, and local governments decide to allocate funds. The City Budget Project highlights the ways in which our city utilizes taxpayer dollars; even if my learners are not of tax-paying age, their parents pay into this system and one day they will too. Even now, they still utilize the services offered by the city. The purpose of this project was to have learners identify institutions they believe are of importance to them, based on their lived experiences and the communities they come from. Students wanted to reallocate funds to health, parks and recreation, the public library, and animal care services. I created this project not because I wanted to replicate my own ideals or push an agenda onto the lives of my learners, but rather to expose them to the different resources and institutions that are funded by our city budget. Some did recognize that over 60% of funds within the general city budget are allocated to police and fire (City of San Antonio, Management & Budget Office, 2023) and raised questions about it. One learner shared "Police and fire receive the majority of money from the city budget over the past four years. I don't think that's right because as they get more money, other things that are just as important lose money." Another shared, "You'd think that with police and fire getting the most money, there wouldn't be as much crime in the city. Something is not right with that."

The City Budget Project and disruptive STEM teaching

As stated at the beginning of this book, disruptive STEM teaching, "centers on the critical examination of [the] intersectional identity of both teacher and learner through pedagogical performances that are interdisciplinary and subversive. Disruptive STEM education assumes a critical stance that informs all practices." Disruptive STEM education in mathematics in particular is of extreme importance, especially for learners who have been most left behind in the subject: learners of color from urban communities. The City Budget Project is one of the ways that I have attempted to disrupt narratives surrounding mathematics as culturally neutral or as a gatekeeper that influences a learner's higher academic and career trajectory (Ukpokodu, 2011; Douglas & Attewell, 2017).

Mathematics education is not just about teaching to a standardized test but rather a human activity grounded in cultural differences that "attempts to describe and understand physical and social phenomena" (Ukpokodu, 2011; Mukhopadhyay & Greer, 2001). I centered my chapter on culturally responsive teaching framed by critical theorist, Geneva Gay, who defines it as using ethnically diverse learners' cultures, experiences, and perspectives as conduits for teaching (Gay, 2002). Gay's work on culturally responsive teaching most aligned with my work as I created the City Budget Project in response to the events of 2020, specifically the government handling of COVID-19 and *Defund the Police* movements following the murder of George Floyd. These events saturated news cycles and social media, as well as the national conversation leading into the 2020 US presidential election. To exclude learners from this conversation, especially since they too experienced the realities of COVID-19 and bore witness to national protests following the murder of George Floyd would be wrong. Students, like adults, must be a part of this conversation, and this informed the creation of my project. Only when knowledge and skills are aligned with learners' lived experiences can education, especially mathematics education, become more engaging for learners of diverse backgrounds, thus producing higher academic achievement within mathematics (Gay, 2000, 2002; Ukpokodu, 2011).

Creating this project was no easy feat. Even though the project comes at the end of the school year after learners have taken their standardized tests, my school expects me to continue teaching learners knowledge, skills, and objectives to reinforce items learned during the past school year that will be necessary for success on their next grade-level standardized test. Disruptive STEM teaching within a content area that is heavily tested is hard, primarily when those metrics determine learner and teacher success for the school year. And yet it is necessary. Disruptive mathematics teaching cannot always be present within my curriculum, or be the driving force of every unit; therefore it must be subversive. This is true now more than ever, as our sociopolitical climate is embroiled in culture wars wherein legislature is drafted to make it harder to educate learners on issues related to social justice.

Educators must take the time to recognize when and where there are windows within a given unit in the curriculum in which disruptive teaching can be embedded – the City Budget Project is an example of that. The City Budget Project is subversive because even though it aligns to state knowledge, skills, and objectives related to units on proportionality and financial literacy, wherein learners are using computational skills, it also asks them to use these same skills through a social justice lens and apply them directly and critically to how our local government utilizes and distributes funding across city institutions. Here too is where this project is interdisciplinary because the state knowledge skills and objectives learners learn not only have worth in terms of getting questions on a standardized test *right* but also have real-world applications. The culminating piece of this project was to have learners write a letter to city officials sharing their findings and detailing whether they believe funding is distributed fairly across city institutions. Here mathematics intersects with justice and the lived experiences of my learners; in this way, mathematics education becomes authentically learner-centered.

I opened this chapter with the question: What do you think of when you hear the word *math*? I believe that it is only fitting to revisit this question here in this chapter's conclusion. My hope is that *math* is more than just about procedural computational skills and following structured rules that build an understanding of mathematical concepts needed for problem-solving. Rather it can be an approach to describe systems at play. Math, just like English and history, can be a subject in which social-political systems can be dissected, explained, and questioned. My work as a disruptive mathematics teacher does not end with the City Budget Project, rather this activity stands as one mathematics teaching model developed during my journey toward becoming a disruptive mathematics teacher. The journey does not end here.

References

Asante, M. K. (1991). The Afrocentric idea in education. *Journal of Negro Education*, *60*, 179. https://doi.org/10.2307/2295608

Britannica, T., & Editors of Encyclopaedia. (2023, April 27). *Pythagorean theorem*. Encyclopedia Britannica. https://www.britannica.com/science/Pythagorean-theorem

Douglas, D., & Attewell, P. (2017). School mathematics as gatekeeper. *He Sociological Quarterly*, *58*(4), 648–669. https://doi.org/10.1080/00380253.2017.1354733

Gay, G. (2000). *Culturally responsive teaching: theory, research, and practice*. 1st ed. Teachers College Press.

Gay, G. (2002). Preparing for culturally responsive teaching. *Journal of Teacher Education*, *53*(2), 106–116. https://doi.org/10.1177/0022487102053002003

Henry, J. S., & Wing, K. (2021). *What does it mean to defund the police?* Cherry Lake Publishing.

Hunter, R., Hunter, J., & Bills, T. (2020). Enacting culturally responsive or socially response-able mathematics education. In C. Nicol, J. A. Q. Q. Xiiem, F. Glanfield & A. J. Dawson (Eds.), *Living culturally responsive mathematics education with/in Indigenous communities* (pp. 137–154). Brill Sense. https://doi.org/10.1163/9789004415768_007

Ladson-Billings, G. (1995). Toward a theory of culturally relevant pedagogy. *American Educational Research Journal*, *32*, 465–491. https://doi.org/10.3102/00028312032003465

Louka, L. (1993, November 15). Girls' learning odyssey. *Tampa Bay Times*. https://www.tampabay.com/archive/1993/11/15/girls-learning-odyssey/

Mensah, F. M. (2021). Culturally relevant and culturally responsive two theories of practice for science teaching. *Science and Children*, *58*(4), 10–13. https://www.nsta.org/science-and-children/science-and-children-marchapril-2021/culturally-relevant-and-culturally

Mukhopadhyay, S., & Greer, B. (2001). Modeling with purpose: Mathematics as a critical tool. In B. Atweh, H. Forgasz & B. Nebres (Eds.), *Sociocultural research on mathematics education: An international perspective* (pp. 295–312). Routledge.

Office of Management & Budget. (2023). *Adopted operating and capital budget*. City of San Antonio.

Paris, D. (2012). Culturally sustaining pedagogy: A needed change in stance, terminology, and practice. *Educational Researcher*, *41*(3), 93–97. https://doi.org/10.
3102/0013189X12441244

Paris, D., & Alim, H. S. (2014). What are we seeking to sustain through culturally
sustaining pedagogy? A loving critique forward. *Harvard Educational Review*,
84(1), 85–100. https://psycnet.apa.org/doi/10.17763/haer.84.1.982l873k2
ht16m77

Paris, D., & Alim, H. S. (2017). *Culturally sustaining pedagogies: Teaching and
learning for justice in a changing world*. Teachers College Press.

Rocheleau, K. J. (1995). *The effects of high school mathematics and science classes on
wages* (Publication No. 56) [Honors Project, Illinois Wesleyan University].
https://digitalcommons.iwu.edu/cgi/viewcontent.cgi?article=1069&context=
econ_honproj

Schwartz, S. (2021, June 11). *Map: Where critical race theory is under attack*.
Education Week Retrieved April 21, 2023, from https://www.edweek.org/
policy-politics/map-where-critical-race-theory-is-under-attack/2021/06

Shor, I. (1992). *Empowering education: Critical teaching for social change*. 1st ed.
University of Chicago Press.

Tate, W. F. (1995). Returning to the root: A culturally relevant approach to mathematics pedagogy. *Theory into Practice*, *34*(3), 166–173. https://doi.org/10.
1080/00405849509543676

Ukpokodu, O. N. (2011). How do I teach mathematics in a culturally responsive
way? Identifying empowering teaching practices. *Multicultural Education*,
19(3), 47–56. https://www.proquest.com/scholarly-journals/how-do-i-teach-
mathematics-culturally-responsive/docview/926980145/se-2?accountid=
10207

Vitale, A. S. (2017). *The end of policing*. Verso Books.

8
ADDRESSING WICKED PROBLEMS IN TEACHER PREPARATION THROUGH EDUCATOR SOLIDARITY

Randa Elbih, Anita Bright, and Margery Gardner

Introduction

This chapter centers teachers as intellectuals who both leverage and generate theory as part of their evolving practice. Aronowitz and Giroux (1985) call for radical intellectuals to protect the democratic nature of schooling and articulate theories of education that ground their work. Radical intellectualism in schools is often threatened by the need to accommodate normalized social practices and further repressed by a lack of institutional support for theorizing endeavors perceived as beyond the scope of teachers' roles and responsibilities (Cochran-Smith & Lytle, 1993). Teacher educators must help strengthen the critical capacities of future teachers so that they can confront oppressive systems and embrace their intellectualism in the face of neoliberal and bureaucratic opposition. Drawing on insights from our various contexts and perspectives, we describe our collaborative process of curriculum development in service of emergent bilingual learners, and our vision for next steps for disruptive STEM work in teacher preparation.

Preparing teacher candidates to appropriately support all learners is an increasingly complex and demanding challenge for teacher educators. Throughout this project, we worked with general education teacher candidates to bolster their capacities to center emergent bilinguals (EBs, sometimes referred to as English learners) in their classrooms through a culturally sustaining pedagogical framework through which educators recognize, center, and uplift the diverse cultural heritage of students. Lesson study is an iterative instructional approach where three educators all eager to focus on

DOI: 10.4324/9781003395782-12

a specific problem of practice, enact one lesson on multiple occasions. One instructor is selected at a time to spearhead the lesson and each opportunity involves feedback and refinement of the lesson. Lesson study emerged as the most appropriate instructional and research approach based on our varied location and context as well as its "comprehensive and well-articulated process for examining practice" (Fernandez, Cannon, & Chokshi, 2003, p. 171). Through lesson study, we developed and revised lesson materials that centered emergent bilingual learners while addressing ocean sustainability and commercial fishing.

From our standpoint, we considered our project to encompass a "wicked problem" (Buchanan, 1992) in teacher education. As mentioned in Chapter 4, Rittel and Webber (1973) first defined a "wicked problem" as a "class of social system problems which are ill-formulated, where the information is confusing, where there are many clients and decision makers with conflicting values, and where the ramifications in the whole system are thoroughly confusing" (Buchanan, 1992, p. 15; West Churchman, 1967). Our "wicked problem" included situating ourselves in the work of curriculum development and instruction across vastly different teacher preparation contexts and confronting our combined inability to fully anticipate the needs of the emergent bilingual learners that we would encounter. Furthermore, understanding the factors associated with our chosen topic – marine environmental sustainability – and conveying these via curriculum and instruction is a wicked problem.

We worked together to reimagine content area instruction in teacher preparation, which is often heavily siloed. Randa approached the project as a social studies educator, Anita with a significant background in Teaching English for Speakers of Other Languages (TESOL), and Margery as a science educator. By centering environmental justice, we each found rich entry points to equip teacher candidates with the necessary capacities to teach a multi-disciplinary truth. This chapter describes our development of materials for a flexible model through which candidates can substitute other issues that are responsive to the composition and interests of their learners.

Centering emergent bilinguals

A growing number of students receive language assistance in P-12 education, 4.9 million students as of 2016 and upwards of 20.2% in some states (National Center for Educational Statistics, 2019). Yet teacher preparation for working with EBs in the general education setting is lagging. Unless the concentration is TESOL or an ELL program, in many teacher preparation settings, general or content area methods courses do not include an explicit or robust focus on meeting the unique needs of EBs (Quintero & Hansen, 2017).

In many contexts within the US educational system, general education teachers have been absolved of responsibility for ensuring EBs have full access to and support in engaging with curricular materials and content. While stand-alone ESOL endorsement programs provide a depth of expertise that is essential, all teacher candidates should have, at minimum, a working knowledge of how best to leverage content-specific pedagogical practices in the service of EBs.

A tension exists within teacher preparation between literacy acquisition skills for EBs and content area knowledge obtainment. Von Esch and Kavanagh (2018) advocate for the need to build capacity among teacher candidates as adaptive experts who can convey learning objectives with proficiency but also maintain a flexible stance in order to meet the needs of their students in a timely and culturally responsive manner. Certainly, the solution is not located in the content-neutral teaching that pervades many English as a New Language (ENL) programs. Lucas (2011) cautions educators that presenting EBs with watered down content results in lower academic expectations and the perpetuation of false narratives about EB ability. We applied a culturally sustaining pedagogical (CSP) framework to expose EB's to rich content area inquiries that are well contextualized and celebrate cultural and ethnic heritage, drawing from each students' unique fund of knowledge (Gonzalez, Moll, & Amanti, 2005).

Paris (2012) initially described culturally sustaining pedagogy as a framework of love with three goals:

1 To see all students as assets who contribute their knowledge to the curriculum and the classroom.
2 To foreground all students' heritage, mainly those marginalized, while considering that culture is dynamic and continuously evolving.
3 Finally, to avoid negative generalizations about a students' cultures when meeting the needs of individual students.

Paris and Alim (2014) assert that "pedagogies can and should teach students to be linguistically and culturally flexible across multiple language varieties and cultural ways of believing and interacting" (p. 12). Further, they highlight the fact that work remains to be done in envisioning and enacting what CSP could look like in practice. By engaging CSP in this cross-institutional lesson study, we highlight learner and family assets to make students' heritage central and actively dismantle and dispel toxic notions about learners, their families, and their cultures. CSP offers a space to celebrate linguistic diversity and center this in our work.

Teacher educators are a vital part of the preparation process and can leverage their position to model EB content area instruction in ways that enrich and sustain linguistically diverse classrooms. Teacher educators can model reflective behaviors and make visible how teachers can refine and adjust lessons according to student feedback (Von Esch & Kavanagh, 2018). Lui (2013) found that teacher preparation programs are able to best support capacity building through early interventions that incorporate multiple sources, including observation, mentoring, and in-service training.

For these reasons, we selected lesson study as a practice that can bring professional behaviors to the surface that support positive engagement with EBs. Extending these ideas, Heineke and Giatsou (2020) found that prioritization of EBs in curriculum and fieldwork in teacher preparation resulted in greater candidate confidence and increased interest in future work with EBs. Our investigation created a viable model with which teacher candidates can meet the unique needs of the expanding population of EBs and multilingual learners, prioritizing critical engagement with the curriculum that privileges students' funds of knowledge and cultural backgrounds.

Through this work, we posed the following questions:

1 How can educators learn from one another through a reflective lesson study process to develop CSP competencies in their own instruction of content areas methods to more readily center EBs?
2 What culturally sustaining curricular and instructional approaches might teacher educators add to their preparation of teacher candidates to equip future teachers for purposeful engagement with EBs?

Research design

Building upon the project team members' unique expertise in science, social studies, and TESOL, we gathered data from our own experiences and interpretations and from teacher candidate participants. We used autoethnography, which combines characteristics of autobiography and ethnography and allows researchers to understand their experiences in connection to their socio-cultural context (Ellis, 2004; Holman Jones, 2005). Lesson study is best when paired with a culturally sustainable framework that centers on the learners, their cultures, and their unique educational needs. This approach allowed for parallel insights on the participation experiences of both teacher candidates and teacher educators.

Using lesson study as a guiding fabric, we felt empowered to develop flexible yet sturdy curricula that can suit a wider array of teacher candidates (Elbih, Miller, Sheldon, & Wilson, 2022). We recognized the vital importance of reflexivity considering that our unique identities shape our research process and outcomes (Lincoln & Guba, 1985). In particular, Margery and Anita identify as White women and Randa as a woman of color and all of us identify as cisgender. At the time of the study, two of us were junior faculty members and one a full professor at our various institutions. We felt ready to enhance our practice as teacher preparers as a means for further professional growth and as an outlet for development of more critical capacities. We conducted this inquiry with space to interrogate our positionality in the process in order to center our solidarity with teacher candidates, as well as students and families who are EBs.

Context

We focused on three teacher preparation program locations; two in the Northeast and one in the Northwest, each with vastly different social contexts but parallel commitments to supporting EB P-12 students and families. The first institution (Institution A), a large public research institution, offers over 40 certifications through multiple pathways. Institution B is a private Catholic institution with several offerings at both graduate and undergraduate levels in education, including a TESOL certificate. Institution C is an independent, coeducational liberal arts university with a very small teacher preparation program and limited EB-focused instruction.

Teacher candidate participants were enrolled in required coursework for upper elementary and/or secondary level certification. The study went through three iterations. For iterations two and three, class participants provided feedback on their experiences engaging with the module. In the final iteration, three participants were selected for one-on-one interviews. Participants are identified by pseudonyms to protect their identities.

Why commercial fishing?

Our goal was to create curricula focused on critical understandings that centered the experiences and knowledge of EBs and served as conduits for empowerment and transformational educational change. We foregrounded EB scaffolds and support in our collaborative decision making when developing focal materials. A primary goal was making content accessible to EBs through an engaging and critical student-focused literacy learning progression that centered on social analysis and systems critique (Parsons & Brown, 2002).

We engaged in much internal debate on the topical focus of the learning module. Each of us proposed several different themes, but we ultimately selected marine environmental sustainability due to its pervasive global nature. The wicked problem of large-scale marine protections is not easily solvable due to the social dimensions involved. Indigenous communities rely on oceans as a source of nutrients and as a means to sustain core cultural practices, and would be impacted by prohibitions on use. 97% of the world's fishers are located in underdeveloped coastal communities and feel scarcity impacts intensely. Fish constitute a large percentage of global protein intake, comprising over a quarter of the diet of people from underdeveloped countries. 200 million jobs are thought to be related to the fishing industry in some capacity as well (Kituyi and Thomson, 2018).

In order to address the compounding issues associated with commercial fishing, we sought out curricular materials that were both interdisciplinary and evocative. We opted to ground our lesson study around the 2021 documentary, *Seaspiracy*, by British filmmaker Ali Tabrizi because the film exposes some of the major issues related to marine health alongside many intersectional societal dimensions. The film provides stunning and visceral visual engagement with the issue of ocean system protections and contains appearances of greats such as scientist and advocate Sylvia Earle. The introduction of documentary as a genre also supports student development of interdisciplinary core knowledge and the power of film as a platform for activism. The documentary is also imperfect, like all teaching resources, and therefore open to student critique and discussion.

First iteration

The first iteration of the project took place with graduate-level teacher candidates at Institution A, Anita's home institution, in summer, 2020. Working together online (due to the global pandemic), Anita assigned her teacher candidates a "deep dive" group activity in which they were called upon to contend with a key issue in education as related to equity and social justice. The central topics in the course were racism, classism, and sexism as they appear in educational contexts, so group topics for the "deep dive" extended to the following eight ideas: ableism, ageism/ adultism, animal rights, cultural appropriation, environmental justice/ environmental racism, fat shaming, linguicism, and the school to prison pipeline. The teacher candidates were asked to dig deeply into their assigned topic, then prepare and facilitate an engaging, workshop-like experience for their classmates to learn from. While this iteration allowed for substantial critical self-reflection, it also brought to light the need for a conceptual anchor through which we could model strategies and develop interdisciplinary discussion. Again, this conceptual anchor we needed was something more specific (rather than a menu of options). For the purposes of this work, we elected to focus on the "wicked problem" as related to commercial overfishing.

Second iteration

Randa, Anita, and Margery all carried out the second iteration in person at institution B in November 2021. The class was an undergraduate course on Teaching in Culturally and Linguistically Diverse Classrooms. Readings for the course took a funnel approach, starting by painting a picture of the diversity in the United States with texts such as Colombo, Cullen, & Lisle (2005) *Rereading America: Cultural Contexts for Critical Thinking & Writing*, then delving more into racial identity questions with Tatum's (2017) *Why Are All the Black Kids Sitting Together in the Cafeteria?: And Other Conversations About Race*, and finally ending with more practical pedagogic resources such as Hollie's (2017) *Culturally and Linguistically Responsive Teaching and Learning – Classroom Practices for Student Success, Grades K-12*.

The teacher candidates were mostly traditional students who were relatively new to the program and therefore hadn't yet completed extensive field experiences. We therefore adjusted the project to address their needs as emerging teacher candidates in the lesson. We developed a lesson plan according to complimentary state and national standards and related objectives noted in the table below (Table 8.1):

TABLE 8.1 National standards and aligned objectives

Standards	Objectives
Social Studies standard: GEO 6–7.3 Explain how cultural patterns and economic decisions influence environments and the daily lives of people.	1 Identify the relationship between fishing practices and ocean health.
Next Generation Science Standard: MS-LS2-4. Construct an argument supported by empirical evidence that changes to physical or biological components of an ecosystem affect populations.	2 Explain how economic decisions such as overfishing result in depletion of fish populations that also have impacts on cultural practices and ecosystem processes.
Connecticut English Language Proficiency Standard: CELP 4: Construct grade appropriate oral and written claims and support them with reasoning and evidence.	3 Debate the issue of overfishing from different perspectives. 4 Share responsibility with other local and global citizens to find solutions to restore the fish populations in the oceans.

The lesson sequence took the following path:

To open the lesson, Anita led the students through a land acknowledgment which honored the homelands of native people from the area. Anita then connected students with land and environment by asking about their relationships with bodies of water, how they engage in self-care, and how they care for the environment. These questions set the stage for our lesson, "Can fishing ever be sustainable?" which represented the overarching compelling question for the lesson.

Randa, Anita, and Margery then divided the body of the lesson into several supporting questions to help students answer the overarching compelling question. The first supporting question was, "What is meant by 'The Tragedy of the Commons'"? Anita started us off with a read aloud of Bang's (1997) book *Common Ground: The Water, Earth, and Air We Share*. The tragedy of the commons is useful for broadly understanding the impact of human resource consumption on the environment. Hardin (1968) argued that increases in human population inevitably result in over-use of public areas that eventually leave the environment depleted beyond repair. "Of course, it is large industrial fisheries that bear responsibility for these effects, not small-scale fisher people" (Micha Rahder, personal communication). Hardin noted that open land spaces were the first to be privatized followed by considerations for pollution outputs post-industrialization. Privatization makes optimal resources accessible only to those in power or with adequate financial capital. Hardin acknowledges how the tragedy of the commons operates within

> the oceans of the world [that] continue to suffer from the survival of the philosophy of the commons. Maritime nations still respond automatically to the shibboleth of the "freedom of the seas." Professing to believe in the 'inexhaustible resources of the oceans,' they bring species after species of fish and whales closer to extinction (p. 1245).

Common Ground: The Water, Earth, and Air We Share brings the tragedy of the commons into focus through visuals and kid-friendly language that makes it accessible to a wide array of audiences and can lead to follow-up discussions around the meanings and messages of the book.

When taking turns, we read in loud, clear voices, at a highly accessible, relaxed pace, which is a strategy to engage emergent bilingual students and bring the class together as one. Students were able to respond to discussion questions and all students agreed with the author's concern that the common resources that humans share should not be exploited or misused. This was an intended result to scaffold learning for the next supporting question: What threats to the ocean and its inhabitants are most pressing?

For this question, we played portions of the *Seaspiracy* (2021) documentary. As the film was too long to show in a class period, we selected certain sections to highlight, primarily focused on key ideas relevant to the main arguments of the film. These clips included a discussion of declining shark populations and their impact on the food chain, fishermen in a tiny boat requesting fish from a large ship, narrative by oceanographer Sylvia Earle, and some debunking of the media narrative as related to the relative impacts of drinking straws vs. fishing nets.

After each clip was played, students were given about two minutes to reflect on what they observed and their layered reactions. Margery and Anita engaged with the students by listening to their conversations and offering perspectives as requested while Randa cued and played the next sections of the video. The students were encouraged to take notes to be able to respond to the supporting question. After watching the video clips, students were asked to respond to three questions in a whole class format:

1 What can we do about this as US citizens and as global citizens?
2 What content knowledge and skill-based goals do we want our students to acquire when teaching this topic?
3 How can we teach about ocean sustainability across different grade levels?

The documentary elicited shock from some of the students, who rarely thought about what happens in the ocean and how the fishing industry is run. To bring clarity to the main conceptual focus of the lesson and allow students to think through the mindset of stakeholders, Randa, Margery, and Anita dedicated the last portion of the class to a role play debate simulation, representing the perspectives of a CEO of a commercial fishing company, an environmentalist, a local fisherperson (non-commercial), a seafood consumer, and judges/evaluators (intended to be as neutral as possible).

Each group was given a few minutes to present their arguments before the panel of judges. The judges then posed questions to each group, seeking to probe for further clarification and general understandings. Finally, the judges decided whose argument was most compelling based on the assembly of evidence.

We were all very surprised to find that the judges ruled in favor of the CEO stakeholder group! They found the CEO's argument that ocean health should be the responsibility of international government agencies, which should maintain it through regulation of fishing and subsequent consumption compelling. The need to generate jobs and global income through commercial fishing also was cited as a viable argument by the class.

This result was concerning because we had anticipated a very different outcome. This group of students were all enrolled in a comparatively expensive private institution of higher education, wherein the economic privileges and/ or lack of criticality of the group were rendered visible as a result of this activity. Early in the lesson, formative responses from the students signaled perspectives that foreshadowed their later viewpoints, mainly in their relationship to water. Their responses suggested that access to clean water sources for daily health as well as recreation was taken for granted and not framed as a privilege or luxury. Even as we realize we do not know the life circumstances of any student, taken as a whole, the demographics of the students in the class represented multiple layers of economic privilege – a form of privilege that may have blurred or obscured the students' abilities to see the issues in the more complex or nuanced ways we had anticipated.

In full recognition of our own biases, we acknowledged that our initial intention behind the module was to develop civic responsibilities and encourage local activism for social and environmental justice. Given the complexity of these ideas, upon reflection, we realized that a more flexible instructional time (Gardner & Tillotson, 2020) for this group of students to explore the multitude of ways in which CEOs and other corporate leadership are responsible for such sweeping environmental impacts, not to mention all of the interconnected social, economic, and political dimensions. However, even as we reckoned with our own worldviews and caution in imposing these worldviews upon the teacher candidates, we applauded their willingness to defend their thinking, even though there was a misalignment with our own lesson goals. We underestimated the far-reaching impacts of capitalism to the point of being perceived by students as natural and thereby normative and not worthy of interrogation. As educators, even as we ourselves are entangled in capitalism, we realized how difficult it can be to dismantle this level of loyalty to "how it is" within a single class meeting. We ponder the ways in which critical perspectives may be better and more holistically incorporated into all of our disciplines, such that teacher candidates may vividly see and understand the ways deeply instantiated systems of power are made manifest across all aspects of being, and particularly within education.

As educators engaged in the construct of disruptive STEM, we were humbled in this response from students (Bright, Acosta, & Parker, 2020), and found a need to sit with the insights shared by them. We felt the overwhelming urge to consider ways to best move forward with the next group of learners, inclusive of our new, shared realizations. To this end, we conceptualized our vision for our third iteration of this project, described in the next section.

Third iteration

We facilitated the third iteration at institution "C" during the last week of the spring semester in April, 2022, using a hybrid format. During the second iteration, we had not assigned students to fully watch the documentary *Seaspiracy*, which, in hindsight, felt like a mistake. We felt as though the students lacked substantive background knowledge to inform their engagement with the questions, which may have been a contributing factor to why the debate outcome transpired the way it did. There may not have been enough buy-in on the significance of commercial overfishing by the institution B students because of this shortcoming. As a result, we made the ambitious pivot to assign the entire documentary to the institution C students as a means to develop greater student awareness of key environmental issues.

The students in this class were a mix of graduate and mostly upper level undergraduates, with many transitioning into an intensive student teaching experience the following semester. While assigning the documentary seemed like a viable solution at the time, it was difficult to motivate students to watch the documentary in its entirety for a variety of reasons, including the explicit images of dolphin, shark and fish murder as well as timing of the lesson in the final weeks of the semester. We also had to adjust our lesson pacing to account for the need to debrief about the film. Time allocation was a struggle for the team due to the ambitious nature of the project. If a longer period of time could be allocated to present these materials, we would have students watch the film in class where debriefing could occur more organically. Since the film is packed with information and is rather fast paced, keying in on the most relevant clips for the lesson would be worthwhile to support meaning-making.

Another shift based on student feedback from institution B was the need to more explicitly include pedagogies that centered EBs. For this reason, we kicked off our debrief of the film by asking students to summarize their reaction to Seaspiracy in a single word. Anita then described how this strategy supports EBs by reducing the anxieties around speaking and offering an opportunity for EBs to echo others if they struggled. She explained it as "one word that expresses what you're learning or feeling right now, and as we speak these, they will become an oral poem." I noticed that my word "capitalism" was also used by another student in the class, and there several other words were repeated as well. Student responses included the words: performative, upsetting, shocking (x2), capitalism (x2), veganism, staggering, frustrating (x2), disbelief, agitated, confusing (x2), tragic, wrenching. During a post-lesson debrief, one student commented specifically on the effectiveness of this strategy because of its ability to disarm students and open space for discussion for everyone, "I liked that we all went around and added one word - it felt pressure-less and it got everyone in the room to say something so that we felt comfortable speaking moving forward" (Taylor, survey, 4/29/22).

The lesson study was embedded in a required course for teacher candidates that focused on asset-based pedagogies. Readings for the course ranged from foundational works like Gonzalez, Moll, and Amanti's (2005) *Funds of Knowledge: Theorizing Practices in Households, Communities, and Classrooms* to contemporary scholarship like Paris and Alim's (2017) *Culturally Sustaining Pedagogies: Teaching and Learning for Justice in a Changing World*. In alignment with the course objectives, we wanted to promote class conversation about the social and cultural implications of the film. This became folded into the debrief through the following prompt: How is the notion of "culture" taken up through the lens of the filmmakers? This prompt sparked almost an immediate contribution from one student who majored in Japanese and had traveled abroad there. She found the film to be far too Western-centric and lacking in counter narrative from Japan citizens who could provide different cultural insights.

This compelling student comment set the tone for most of the remainder of the lesson and due to its fierce indictment of the documentary left the remainder of the class seeming entirely skeptical of the film. Melanie picked up on the fear factor of the film and the way the visuals portray the issue as dire:

> I'm not tuned into the environmental news and question the validity. …I usually don't take the time to question. Some context readings would be helpful to affirm the research and I feel that it pulled me in a way that was scary.

For her, the decision to include unsettling imagery of the impact of commercial fishing seemed to evoke further skepticism. The gruesome, visceral scenes of harvests of fish and dolphins actually clouded, in her mind, some of the key ideas, and reduced the effectiveness of the filmmaker in delivering a powerful central argument about the harms of commercial fishing.

Both Melanie and another student thought that the film needed a balance point, suggesting the incorporation of articles either to support its claims or to illustrate a multicultural perspective on the topic. "I think that the video leaves some things about culture out (I think it pushes a single side) so maybe having a counter article about reactions from those bodies in the mix of the lesson would be good." We appreciated the students' attention to multiple perspectives yet wonder if this heightened criticality created barriers to investing in learning some of the core issues. An avoidance strategy emerged that allowed for personal distance from complicity as they processed the graveness of the issue. We certainly agree with the need to decenter the White, Eurocentric gaze of the film but this lack of representation didn't call into question (for us) the central claims about the environmental impact of the fishing industrial complex. This observation calls to attention the root cause of such hesitancy as part of the post-Trump, fake news era in the United States.

The lesson then deviated into breakouts led by each team member to highlight a particular teaching practice that targeted the needs of EBs through the lenses of literacy, mathematics, and discussion. Anita started us off again with a read aloud of Bang's (1997) book *Common Ground: The Water, Earth, and Air We Share.* At institution B, we each took turns reading the text aloud, but it was difficult for all the students to see the striking imagery. For that reason, we opted for a read aloud of the book by an educator on YouTube so that it could be enlarged on the white board. The read aloud with the clearest vocals and best visual display also had background music. This created a tension for our team because the background music could be distracting for EB students and therefore actually act as a barrier for processing the contents of the book. After some debate, we decided that we would go ahead and use the read aloud but introduce the material as imperfect. In hindsight and with more time, we would have created this read aloud video in house in order to have the material align more closely with our aspirations.

Through transparency about our thought process, our lesson study team thought that the common struggle to identify quality curricular materials could be shared and folded into an honest discussion about practice. The mini lesson that followed the read aloud included a couple of recommended practices that support EBs' reading comprehension. These include predicting the text prior to reading by analyzing the cover page, previewing vocabulary that may be unfamiliar, including idioms, leveraging text and graphic features throughout the reading, and word tracking while reading. Anita sought out feedback on the book from the class and pointed out gendered language such as "fishermen" rather than "fisherpeople." During a short debrief about the book, graduate student Melanie noted that the pictures were very busy and EBs would require processing time to unpack the visuals. Another student participant commented that the read aloud was one of the moments in the lesson that stood out as most engaging, "I also enjoyed the different forms of media, the reading of the online book added to the conversation on linguistic diversity while connecting to the Seaspiracy film." What students tended to notice and appreciate from this portion of the lesson was Anita's attention to clear diction and slower talking pace. As an additional factor to consider, Anita and Randa were joining the class via Zoom. This instructional tactic best supported remote instruction when transmission can be distorted or delayed while also demonstrating the need to honor processing time for EBs.

Next, Margery walked the class through a demonstration on energy return-on-investment (EROI) that she had learned working with Dr. Rick Beal at the State University of New York College of Environmental Science and Forestry well over a decade previously. The activity includes a math sheet with a series of multiplication problems that range from single-digit to five-digit numbers (see Appendix). The class is instructed to take one minute to complete as many of the problems as they can. Once the minute is over, the students count how many problems they've successfully completed and record this number. They continue for two more rounds and in the last round they have to work collaboratively with a partner. When the task is complete, the students can recognize a clear declining trend in the number of problems successfully completed during each round. This simulation is used to draw direct parallels to resource consumption, where humans first extract the most desirable and readily available resources followed by diminishing returns. Eventually, the amount of effort put into extraction equals the resources put in. As these sources become depleted, there is a need to seek out alternatives that are more challenging to gather. This concept offers a framework for understanding a range of phenomena including organismal energetics, hunter/gatherer foraging strategies, and even economic choices associated with fuels such as oil production. It was Margery's hope that students would be able to connect the activity directly to the film in order to understand why commercial fishing outfits go to such great lengths to obtain their harvests at substantial costs.

Many of the student participants immediately picked up on the fact that the simulation activity resembled timed fluency programs such as "Mad Minute." This realization triggered groans by most when first introduced. Margery had to comfort the class by letting them know that their work would only be self-evaluated. During the activity debrief, Margery pointed out the specific strategies that best serve EBs, including short and clear directives, little to no extraneous text, and an opportunity to partner with a peer before whole group discussion. In post-lesson reflections, student participants found this activity quite engaging, but this reaction could be in part due to the fact that this portion was completed in person with their own instructor with whom they felt most comfortable. Matthew, a junior geology major, determined his own connection to the EROI principle that grounded his work in the earth sciences and as a future secondary teacher.

> I liked the math portion a lot and how to relate more interdisciplinary subjects. It made me think about partial crystallization in geology, there can be layers based on the cooling patterns. Math and vocab could be included into more science activities. It felt nostalgic. I remember flipping over the sheet in fifth grade. For vocab, having very easy to define as well as more abstract concepts.
>
> *(Interview, 5/5/22)*

With time waning in the lesson, Randa then modeled the final breakout portion that focused on building multiple perspectives. This was another pivot from our prior iteration, where we introduced a debate activity but didn't provide adequate time to truly unpack the dimensionality of the issue. Randa discussed how the inclusion of multiple perspectives can allow for the development of a classroom community, minimize harmful hierarchies, offer opportunities for self-reflection, and allow students to try out different methods of sharing information. Randa's discussion included a short definition of the terms capitalism and neoliberalism as well as a comparison of the two. Taylor noted how appreciative she was of this instructional move and noted that it would be very important to incorporate explicit definitions of key terms with EBs. Margery was a bit surprised by this comment because she believed that the class, given its exposure to critical theories of education, would have a strong handle on these terms. It was a clear learning moment for her that there is a need to provide instructional space to offer common language for students at all levels and to avoid making assumptions in these areas.

Overall, the lesson study supported students' emerging understanding of teaching pedagogies by acting as a way for the class to actively see how certain concepts are put into practice. For instance, Margery assigned Budd-Rowe's (1986) classic study on wait time as part of her course. In this construct of "wait time," there are two key aspects of waiting involved. First, the facilitator poses a prompt or question, and allows substantive time for students to think and process. Next, as students respond to the prompt or question, the facilitator also waits to affirm, further probe, or move forward from the reply. The students immediately picked up on our team's use of wait time and complemented our use of the concept. Out of eight student responses post-lesson, five noted the use of wait time by instructors. In many ways, the lesson study was a snapshot of what students were encountering through their coursework as evolving teacher candidates.

For this final lesson study round, as in the prior version, the team overstuffed the lesson with a multitude of plans that weren't actualized. We then had to prioritize certain components over others in real time. For example, we also wanted to introduce an application phase where the class would create an educational materials tab for the *Seaspiracy* website and reach out to the creators to inquire about the possibility of including lesson plans and other items for teachers who plan to view the documentary. Unfortunately, this section of the planned lesson study curriculum never ended up happening. The time and labor to collaborate on developing the lesson focus and goals alone took several sessions. The additional time to travel to each site and conduct the lesson study was also prohibitive and could only be completed with all team members in person on one occasion. We all agreed that it would be ideal to spend more time on the topic of commercial fishing and extend the lesson study beyond the single session but with other professional obligations in play this became nearly impossible to achieve, especially given the structures of higher education and the pressures associated with teacher preparation to complete various state and national requirements.

The process of lesson study had deep impacts on our perspectives on teaching. It helped to unlock different resources and offered an open forum for reinventing practice through collaboration. These opportunities are not readily available for teacher educators who work within confines that are more outcome based than process oriented. It also required bravery and vulnerability to enact lessons in front of esteemed peers. Margery noted that she became better at both giving and receiving constructive feedback as a result of the experience and we would all be interested in engaging together again in lesson study.

The lesson study process also brought into focus the fact that teaching and learning are human endeavors that can never be fully mastered. This was evidenced by the team's struggle to land on a topic that was critical while allowing for smooth interdisciplinary connections to naturally form. Melanie, a graduate student with prior teaching experience, critiqued the use of oceans and fishing as a central fixture because of concerns that this wasn't applicable to many of the students she was familiar with working in an urban setting. "Seafood is expensive and students might not have cultural connections. Water contamination is an issue in urban areas. Maybe there's another film that could do that" (Interview, 5/2/22). Taylor, an undergraduate senior, also noted the importance of context and suggested extension activities that would allow students to research a related topic and then share their findings with the class, possibly through a visual display such as a poster.

Connecting to disruptive STEM

Through this lesson study, we aimed to develop a curriculum to shine a critical light on the harms associated with commercial marine harvests as an intersectional fight for justice. As noted previously, we continue to strive to highlight the distinctions between individual culpability on the small-scale, in contrast to the tremendous harms inflicted by capitalist mega-corporations. We note how big business, as well as other entities with power, such as educational systems, demonize individuals, and particularly those already experiencing marginalization, in ways that further their oppression. As such, we seek to provide opportunities to unpack the wicked problems around societal discourses related to these power imbalances and structures. We hope to support learning and awareness across an array of contexts, from a religious affiliated undergraduate institution, to an elite liberal arts institution, to a more liberal leaning large public university. We highlight the need to understand the origins and impact of our food source choices. We prioritize listening to indigenous communities to create sustainable fishing futures. Running parallel to this pursuit is the desire to center EBs and envelop the module with the warm embrace of culturally sustaining pedagogy. While certain aspects of this challenge were met more successfully than others, we believe that striving for such lofty goals is worthwhile and necessary to confront this and other wicked problems in education.

> The implications for teacher education, then, (are) clear. Teacher educators (need) to nurture the capacity for joy and wonder in aspiring teachers. We (need) to invite into our own classrooms at the university both messy materials and messy social contradictions. We (need) to take time to study critical text, but in a context that left space for playing out the socially reconstructive potential in what we came to understand.
>
> *(Regenspan, 2002, p. 15)*

We each brought in the perspective of our region and would not have had the chance to experience this otherwise. Initially, we all thought about the lesson from our own unique perspective. We were able to coalesce these different facets to unlock new concepts and ideas. Using the *Seaspiracy* documentary as the key text, we were able to collectively subvert what science, social studies, or teaching ESOL should look like. We were able to evolve professionally as teachers due to this rich environment where adaptability was normalized and highly celebrated. We engaged in cross-institutional sharing in ways that were productive and long lasting to us personally. We noticed that civic mindedness was a binding thread that we all sought to advance through solidarity and a commitment to justice.

We all have different relationships with our food sources, specifically fish and seafood. Anita is a vegetarian and therefore has less connection with fish as a food source. Randa and Margery still eat fish and seafood and therefore have to reconcile their choices with the associated environmental impacts. Positioning ourselves as part of a system is also illustrative of our work as critical teacher preparers. Through the use of lesson study as a structure, we were able to see ourselves and our teacher preparation work as bigger than our singular perspectives, gaining fresh insights to how we approach the teaching process as well as our enactment. The vulnerability of teaching in front of respected colleagues and inviting open and honest feedback from our students allowed for unguarded growth and a diffusion of hierarchies.

References

Aronowitz, S., & Giroux, H. (1985). Radical education and transformative intellectuals. *Canadian Journal of Political and Social Theory*, *9*(3), 48–63.

Bang, M. (1997). *Common ground: The water, earth, and air we share*. Scholastic Inc.

Bright, A., Acosta, S., & Parker, B. (2020). Humility matters: Interrogating our positionality, power, and privilege through collaboration. In S. Keengwe (Ed.), *Handbook of research on diversity and social justice in higher education*. IGI Press.

Buchanan, R. (1992). Wicked problems in design thinking. *Design Issues*, *8*(2), 5–21.

Cochran-Smith, M., & Lytle, S. L. (Eds.). (1993). *Inside/outside: Teacher research and knowledge*. Teachers College Press.

Colombo, C., Cullen, R. & Lisle, B. (2005) *Rereading America: Cultural contexts for critical thinking & writing*. Bedford Books of St. Martin's Press.

Elbih, R., Miller, E., Sheldon, G., & Wilson, M. (2022). Integrating social studies education with mathematics: Pre-service teachers' use of the pyramids of Giza to plan a STEM lesson. *Current Issues in Middle Level Education*, 26(2), 1–8.

Ellis, C. (2004). *The ethnographic I: A methodological novel about autoethnography*. AltaMira Press.

Fernandez, C., Cannon, J., & Chokshi, S. (2003). A US–Japan lesson study collaboration reveals critical lenses for examining practice. *Teaching and Teacher Education*, *19*(2), 171–185.

Gardner, M. A., & Tillotson, J. W. (2020). Explorations of an integrated STEM middle school classroom: Understanding spatial and temporal possibilities for collective teaching. *International Journal of Science Education*, *42*(11), 1895–1914.

Gonzalez, N., Moll, L., & Amanti, C. (2005). (Eds.), *Funds of knowledge: Theorizing practices in households, communities, and classrooms*. Routledge.

Hardin, G. (1968). The tragedy of the commons. *Science*, *162*, 1243–1248.

Heineke, A. J., & Giatsou, E. (2020). Learning from students, teachers, and schools: Field-based teacher education for emergent bilingual learners. *Journal of Teacher Education*, *71*(1), 148–161.

Hollie, S. (2017). *Culturally and linguistically responsive teaching and learning: Classroom practices for student success*. Teacher Created Materials.

Holman Jones, S. (2005). Autoethnography: Making the personal political. In Norman K. Denzin & Yvonna S. Lincoln (Eds.), *Handbook of qualitative research* (pp.763–791). Sage.

Kituyi, M. & Thomson, P. (July 13, 2018) 90% of fish stocks are used up – fisheries subsidies must stop. *United Nations Conference for Trade and Development*. https://unctad.org/news/90-fish-stocks-are-used-fisheries-subsidies-must-stop

Lincoln, YS. & Guba, EG. (1985). *Naturalistic inquiry*. Sage Publications.

Liu, S. (2013). Pedagogical content knowledge: A case study of ESL teacher educator. *English Language Teaching*, *6*(7), 128–138.

Lucas, T. (2011). *Teacher preparation for linguistically diverse classrooms*. Routledge.

National Center for Education Statistics (NCES) (2019). English Language Learners in public schools. Retrieved from https://nces.ed.gov/programs/coe/indicator_cgf.asp

Paris, D. (2012). Culturally sustaining pedagogy: A needed change in stance, terminology, and practice. *Educational Researcher*, *41*(3), 93–97.

Paris, D., & Alim, H. S. (2014). What are we seeking to sustain through culturally sustaining pedagogy? A loving critique forward. *Harvard Educational Review*, *84*(1), 85–100.

Paris, D., & Alim, H. S. (Eds.). (2017). *Culturally sustaining pedagogies: Teaching and learning for justice in a changing world*. Teachers College Press.

Parsons, R., & Brown, K. (2002). *Teacher as reflective practitioner and action researcher*. Wadsworth/ Thomas.

Quintero, D., & Hansen, M. (2017, June 2). English learners and the growing need for qualified teachers. In *Brookings*. Retrieved from https://www.brookings.edu/blog/brown-center-chalkboard/2017/06/02/english-learners-and-the-growing-need-for-qualified-teachers/.

Regenspan, B. (2002). *Parallel practices: Social justice-focused teacher education and the elementary school classroom. Counterpoints*. Peter Lang Publishing, Inc.

Rittel, H. W. J., & Webber, M. M. (1973). Dilemmas in a general theory of planning. *Policy Sciences, 4*(2), 155–169. https://doi.org/10.1007/BF01405730

Rowe, M. B. (1986). Wait time: Slowing down may be a way of speeding up! *Journal of Teacher Education, 37*(1), 43–50.

Tabrizi, A. (2021). *Seapiracy*. "Film" A.U.M. Films/ Disrupt Studios.

Tatum, B. D. (2017). *Why are all the Black kids sitting together in the cafeteria?: And other conversations about race*. Hachette UK.

Von Esch, K. S., & Kavanagh, S. S. (2018). Preparing mainstream classroom teachers of English learner students: Grounding practice-based designs for teacher learning in theories of adaptive expertise development. *Journal of Teacher Education, 69*(3), 239–251.

West Churchman, C. (December, 1967). Wicked problems. *Management Science, 4*(14), B-141–142.

9

LEARNING WITH AN UNPREPARED MIND IS LIKE WALKING INTO A FOREST WITHOUT A MAP

Developing reflective meditation practices in math

Payal Patel

Butterflies before starting new puzzles. Panic at the possibility of failure. Dread at the thought of new math adventures. Rigid thought patterns of "I'm not good at math." Habits of using each opportunity to "confirm how good or bad I am at this." All of these are pervasive in many students' experiences of studying math.

As a graduate student of mathematics education at the University of Pennsylvania Graduate School of Education, I began to notice the patterns of nervousness and unhealthy thinking about learning math among students. As I proceeded further into my career in education, teaching in a combination of rural, suburban, urban, public, and private schools, these patterns came up again and again regardless of students' age, gender, socioeconomic status, state, or school. Some major barriers to learning mathematics came to light. I observed an obsession about the end result of a problem or final grades rather than following the journey of learning, which inevitably involves discomfort, irritation, excitement, annoyance, complete satisfaction, and joy. I noticed a struggle to focus while working, fixed mindsets about one's abilities in math, and varying levels of math anxiety.

The goal of teaching math is not to merely feed students content or cultivate disconnected individual skills in students, and it does not revolve around a single authority who holds absolute knowledge. This would in fact diminish and negate the possibility of the profound journey of learning that takes a child into adulthood and beyond. Instead, we should cultivate students' ability to experience an immersion of the whole self in a landscape of deep intellectual and intense emotional experience. This potential is not just held in the mind but also occurs in the body – without a body, none of us

DOI: 10.4324/9781003395782-13

or our journeys in learning would exist. Every child, regardless of race, gender, age, sexual orientation, and the whole host of differences that exist, has an unimaginable capacity to learn. My use of mindfulness and reflective meditation seeks to inspire and bring out hidden gems latent in students.

Lisa Delpit's work on "infinite capacity" (2012) discusses similar ideas in relation to traditional African education as "a quest...to eliminate foolishness. It is foolishness that keeps a person from learning... Africans assume that people have the mental capacity to achieve, but they are concerned about the 'software' that allows brilliant people to misuse the capacity" (p. 28). She asserts that it is beyond "doubt that all humans are capable of learning" (p. 29). As educators, we must not only believe in but also continuously make use of practices that again and again implicitly let children know that they truly have infinite capacity.

In exploring various meditation exercises with our students, as I discuss in this chapter, this immense potential in a child has the possibility to surface. After all, meditation is an antidote to the "mind–body split" referred to in bell hooks' (1994) work in *Teaching to Transgress*. The math classroom is not a place where students leave their emotional reactions, feelings in their bodies, and personal history at the door. It is not just a group of intellects sitting in a classroom focused on the last step and final answer. Processing emotions and the preceding and succeeding feelings in the body throughout the journey of learning math must be addressed and navigated. The wholeness with which each child views themselves is important in sustaining the journey of learning mathematics. This is why I embarked on creating variants of meditation and joining my students' journeys using meditation exercises that help personify emotions, develop playful curiosity, and create visualizations to process a mind–body experience. As hooks eloquently describes, the classroom regardless of its "limitations...[is] a location of possibility" (p. 207). She beautifully describes learning as a "place where paradise can be created" (p. 207). What better place to cultivate paradise than in our whole selves, perhaps by implementing a reflective meditation practice for our students where mundane intellectual activities can meet the places within us that are curious, spirited, and free.

My journey into understanding these struggles, however, did not begin in the field of education. Graduating as salutatorian of South Brunswick High School in the Princeton, NJ, area, I was accepted to a joint BA/MD program at Boston University. While in college and medical school, I spent a significant amount of time mentoring and tutoring high school and college students in STEM. After a change in my career path during medical school, I pursued the study of mathematics. Along this journey, I discovered my deep connection to the art of teaching. Teaching and engaging in growth and learning would send me into a flow state: invigorated, curious, and immersed in the art. During this discovery, it also became very apparent to me that the mind could either hamper or help one learn and grow in every moment. There is an immeasurable latent power in the mind that can be tapped to navigate our way through every moment, and every moment holds the potential for immense learning.

As I proceeded to teach in classrooms, repeated observations of students struggling with mathematics was concerning to me as a mathematics educator. As the world grows more complex, no field stands by itself and mastery of mathematics from a young age makes a huge difference in the lives of students in a complex interconnected world. In their role as global citizens, students are also, more than ever before, presented with constant streams of numbers in the context of scientific debates, economic changes, and socio-economic inequalities. In order to make informed decisions that contribute to progress in society and the world as a whole, students' fluency in logically analyzing numbers and understanding patterns is indispensable.

This leads to an even more urgent concern, as an educator. How will my students voice their opinions and stand up for their stances to support effective democratic changes if mathematics is frightening and a source of confusion? More importantly, this struggle uncovered students' approach to learning when a situation turned difficult. It seemed that most students struggled with the art of learning from each moment, which requires sitting with frustration, persevering before challenges, redirecting focus to the present moment, and letting go of expectations of how and when an end result should appear (if it does) (Figure 9.1).

FIGURE 9.1   Diagram of student struggling.

I began wondering what could be done to address this issue. No matter where you begin in the cycle, it leads to another obstacle to learning and thriving in math, and the more vicious the cycle grows, the bigger and bigger the problem gets – likening itself to the snowball effect. For example, frustration leads to lack of engagement, which then leads to a distracted mind. This leads to stress as learners feel lost on a journey that seems to have overtaken them. It triggers a downward spike in confidence, and the cycle continues with greater momentum as time goes by.

As I wondered about this situation, my mind organically gravitated toward my childhood. My parents immigrated from India, and, as an Indian American, I was fortunate to be exposed to and immersed in ancient Indian philosophy on harnessing the power of the mind–body connection through numerous techniques and how to open the mind up to accept and surrender to infinite possibilities that exist in every moment. Having experienced the impact the ancient practices had on my own growth, which relies on one's ability to learn from every experience and from moment-to-moment, I was inspired to use the ancient techniques of "mindfulness" and what I call "reflective meditations." I first experimented with these techniques myself while learning of stories from the Mahabharata, the longest epic ever to be recorded, in which ideas of the mind are embedded in complex stories within stories. Before stressful situations or academic assessment in high school, I would sit down in my bedroom and use my breath to infuse calm through my body. As I inhaled, I would imagine a stream of cool white milk, a strong symbol in my heritage, entering my belly and cooling any sources of reactivity. As I exhaled, I could seep further into a coolness that entered my belly. It would refresh my being. Every inhale took me back to being a small, free-spirited child, living life spontaneously without worry in the face of peer pressure and examinations, all sources of judgment. This is the most concrete example I vividly remember from my childhood.

What is mindfulness?

Mindfulness is often linked to phrases such as "be here, now" or "being present." It is the practice of bringing one's attention to the present moment using the breath, taste, sounds, or visual field, or other objects of focus as anchors. In an age of over-stimulation by unlimited choices, ads, quickly sprouting innovations, quick and easy processed foods filled with sodium and sugars, the internet, screens, and other devices, mindfulness offers an antidote to forgetting this unique special moment, the unraveling now, that will never return to us.

In this practice, we peer into ourselves or the outside world as if we are seeing it for the first time, with curiosity and non-judgment, much like a newborn. It is said that babies and very young kids don't differentiate humans based on the differences that tragically pull us apart as adults, like socioeconomic status, race, ethnicity, language, or clothing styles. And so, we practice "beginner's mind," observing like newborns. While mindfulness sounds easy, it is called a practice for the very reason we call anything a practice – it has to be cultivated, or more appropriately, re-cultivated. I do believe that all of us were born with mindfulness at our fingertips. Over time, we get assimilated into a world that is ever-changing and we get swept up in the speed of life. It then becomes important to exercise patience, as cultivating mindfulness is not as easy as clicking on a Google link or swiping a credit card. This inner work simultaneously takes a lot of effort and patience.

As we actively pay attention to the present moment, our minds naturally wander. It is the nature of the mind to wander and, if not tended to, it will scatter without direction. We think of the past – analyzing our past experiences, trying to figure out what went wrong and what went right – or dive into the future – excited or nervous about the unknown. Then there is the realm of daydreaming where the creative parts of our mind construct stories and epics. To bring our attention back to now, we engage in letting go of thoughts of the past, future, and self-invented stories. Letting go is often compared to watching the passing of clouds. Treating distractions as moving clouds, we allow them to float by at their own pace, bringing the mind back to the chosen anchor.

Mindfulness endows the mind with quietude and focus. Imagine a mind whose normal frequency is calm and even. When an intense situation takes place, the mind acts like a pool of water in which a rock has been thrown – ripples form in all directions. After some time, the mind returns to its original even-minded state.

After practicing mindfulness for some time with my eyes closed and then slowly engaging with the practice in my everyday experiences, I noticed how I was much more open to opposing ideas and less judgmental of new situations. I began using the age-old wisdom of patience, beginner's mind (i.e., experiencing the moment as if for the first time), non-judgment, and being here, now in my daily life. I was amazed by the decrease in my nervousness in intense situations and increased ability to sit "with not" knowing the answer to questions and frustrations. These personal experiences inspired me to implement mindfulness in my mathematics classrooms in the hope that my students would find it helpful in journeying through the mathematics puzzles and concepts, as well as in their lives now and in the future.

Mindfulness-inspired "reflective meditations"

Mindfulness prepares the mind to focus and sit with the discomfort of learning and change. As I explored mindfulness techniques in my classroom, they began flourishing into exercises that were more visual, metaphorical, and reflective in nature. It is not uncommon for such techniques to complement mindfulness practices in ancient practices or in modern adaptations. So, it seemed more than a coincidence that they naturally followed the implementation of mindfulness. I developed variants of reflective meditations to specifically facilitate the journey of learning by building perseverance, collaboration, curiosity, self-control, joy, and metacognition, among many other qualities of mind. I construct visualizations, stories, and contemplations to equip learners with messages that will support them through the obstacles in the profound journey of learning in every moment.

For example, as you will see in a script later in this chapter, I use visualizations to help students experience the nature of the mind, which is vast, deep, and ever-changing, like an ocean. We would meditate together on the image of an ocean that represents the mind and then dive into the ocean of the mind only to realize that the waves, representing our thoughts and emotions, brought us soaring high and crashing low. It was quite exhausting, so we would then practice stepping out of the ocean of our minds and watching its waves as we lay on the shore of a sunny beach. Wow! How much more grounding! Students pictured thoughts and feelings as small and large waves, respectively, which created ripple effects on us. Watching as an outside observer was so much easier. Following exercises like this, as we transition to learning, students use these concepts to resist being dragged down by fears of not understanding right away.

Over time, I constructed numerous exercises developing our ability to embrace our personal and sometimes elusive journeys of learning. These range from visualizations to conversations with personified aspects of learning. My students learned to befriend "Struggle," personified as a living being. If used artistically and in the right way, this ageless practice of meditating on tangible objects or images to access sources of clarity, self-control, fearlessness, metacognition, self-care, empathy, and purpose can help us through the tumultuous ride of the learning process and create strong minds.

Implementing mindfulness in the mathematics classroom

I carried out mindfulness practices in the classroom with over 100 middle school and first-year high-school math students in an urban setting. I started out with brief mindfulness exercises and collected anonymous feedback from students on feasibility, focus, and well-being. When asked about feasibility, students' responses were positive, with the majority of students (80%, a number that increased with additional exercises) stating that they would like to do the exercises again. When asked how they felt after the exercise, about half of the students noted increased feelings of well-being and about one-third noted feeling more focused. When asked about any differences in doing classwork pre- and post-mindfulness, about 80% of students noted an improvement in focus. These results were critical as they offered a glimpse into students' responses to mindfulness.

As with any new exercise, when I first began to use it in the 2010s, mindfulness was not as mainstream in schools as it is today and was not common at most of the schools I taught at. I realized that the introduction would be quite important. If you have used mindfulness or other meditation techniques in the classroom, then you have experienced curiosity or possible nervousness in students' reactions during the first iteration. If you have not tried these types of techniques but would like to, I offer a glimpse of what the first day of meditation in my mathematics classes looks like. This is a guide that can be modified based on your students. As you proceed through the chapter, you will also find a note on navigating the exercises online during remote learning.

The first implementation

To provide a picture of how reflective meditation may look when it is first introduced, I begin with a narrativization of my work with middle schoolers based on an amalgamation of stories of these introductions over the years, followed by the first script of this chapter.

The bell for period 3 is about to ring, and I am excited to try my first meditation of the year. It's October now and I have a better understanding of the individual personalities and group dynamics in my classes. I am very excited and curious to see how this year's students react to mindfulness and reflective meditations. My aim is to promote a culture of deep reflection and cultivate a more joyful learning environment in the classroom.

After writing a challenge problem aimed at building pattern-recognition skills on the board that connects the concepts in our current unit to previous ones, I set up my go-to background music to play as students run, walk, and sometimes spring into the classroom onto a friend. The energy is so high. I look at the expressions on all of my students' faces – some are smiling at each other, others looking contemplative as they open up their binders, a few students are looking down lost in thought. One student runs up to me saying, "I forgot my pencil and my binder … and everything at home." I point to the bucket of utensils and extra packets. A few students are peering over each other's work, "Mrs. Patel, we don't get #5 on the homework." They are scrambling toward me showing their papers and the methods they attempted as I note the patterns of thinking evident in their work on paper.

The bell rings. From the energy of the classroom, which seems quite abundant now as the nervousness of September has worn off, I can tell that it indeed has turned out to be a perfect day to introduce the first exercise of the year.

Me:	*Good Morning everyone! Some of you have asked me about a problem from last night, we will explore it together in addition to any other questions you have.*
	Today, before we start math problems, I want to do something very special that I have found has helped me when I am nervous, cannot focus, or just simply need to unwind. It helps me feel grounded. We are going to try an exercise called mindfulness. We feed and exercise our bodies and intellect, and it is equally important to take time to nurture the rest of our selves.
Girl sitting in the back of the class:	*"What do you mean by intellect?"*
Me:	*The intellect is a part of the mind that is associated with problem-solving, rationally comparing two situations, and academic pursuits. The intellect is immersed in highly analytical activity – in schools, we grow our intellectual abilities when we solve math problems, learn grammar, proofread paragraphs and essays, and hypothesize and perform scientific experiments. But this is only one part of us.*
Boy calls out:	*So, like what else is there?*
Me:	*We have a vast world that lives within ourselves and each part of it affects our ability to experience each moment, including every moment of learning. We have emotions, feelings, and random thoughts – you can think of this as the "heart of the mind." Growing up, I was told nothing is as fast as the mind – not even light. If in disarray, the mind can generate tremendous amounts of*

> *scattered energies in the form of racing,*
> *depressing, exciting, or silly persistent*
> *thoughts. It can even make us think and*
> *react in such silly ways.*

I understand that a few of my thoughts may not make complete sense to the young students right now, but I imagine that these ideas will form seeds in their minds. I watch the students' faces as some look at me with curious eyes and raised eyebrows.

An excited student: Cool, so is this mindful stuff like yoga and meditation or
something?

He closes his eyes and makes the sound of "Oooooommmm" as we all laugh and enjoy his humor.

Me: Yes! It is closely related. In fact, all of these practices fall under the enor-
mous tree of yoga. Like little sprouts off of the main shoot. So, ready
everyone!

I see students nodding and excited, while others are looking back at me wondering what all of this is.

Me: I am going to ask you all to find a comfortable position. It is beneficial to
sit up right as this will keep your nervous system and mind alert. The
purpose of the activity is not to just stop doing "stuff" or empty ourselves of
thoughts and feelings. What I am saying will make more sense once we
practice a few times. Ok, and here we go – I have some music too!

Students look at each other, some smiling and some giggling. Others are excited about the calm music as it starts. I begin to read from the script.

Script I

> *If you are wearing glasses you may take them off and place them on your*
> *desk.*
> *Begin by sitting upright, drawing the shoulders back if they are slouching*
> *forward. I will also be doing the same.*

[While you can take a peek to check that students are doing ok, it's comforting to students to not have eyes looking at them.

If you feel this may not work the first time, feel free to keep your eyes open but maybe look down, at your screen playing the music, or at your script.]

Find a comfortable position and close your eyes.
If you find it difficult to close your eyes, lower your gaze until your eyes find a still unmoving point in front of or on your desk.

[I learned this when I worked with students who had dealt with trauma and were scared of closing their eyes.]

Bring your attention to your jaws. Stretch out your mouth wide and close it, allowing the jaws to come to a relaxed position.
Now notice if you might be wrinkling your forehead.
Raise your eyebrows and drop them, imagining a wave of relaxation rippling out from the center of your forehead towards the temples.
Next, take a deep breath in and imagine the breath entering your body like a cool stream of water dropping right into your belly.
As you breathe out, your breath arises from deep within your belly and travels through your chest and finally through the tip of your nose and into the world again.
On the next inhale, imagine your breath flowing down your neck and into your shoulders, circulating around your shoulders, as it collects tension, exhaustion, stiffness, and pain.
On the next exhale, imagine that the tension, pain, exhaustion your breath has collected on the inhale travels back out of your body and into the air around you, disappearing forever.

[A few kids may open one or both eyes, look around, look back at me or each other and smile, and close their eyes again. Most kids will be sitting with eyes closed. One student who may have not gotten enough sleep might be truly basking in this exercise, possibly dozing off for 30 seconds. This is normal and, at the end of the exercise, I remind students of how the exercise allows the body to communicate with the mind and tell them about what they could do to take care of themselves.]

Bring your attention back to your breath as you notice the breath enter through the tip of your nose.
It feels cool like a stream of water on a warm summer day and flows down to the back of your mouth, through the chest, and drops into the belly.
Good.

[Students will feel supported as you say "good" in a very gentle manner – not judgmentally – much like you would to a very young child.]

As you exhale, the breath re-arises from deep within you and follows the pathway back out through the tip of your nose.

Continue doing this, following the stream of the breath in and out of the body.

[Pause for 15 seconds.]

Notice if your mind has wandered into telling stories or thinking about the past or the future.

This will happen a million times, it is natural and normal, and every time this happens, bring your attention back to its focus on the breath, even the million and first time.

Bring your mind back to the tip of your nose where the breath enters and imagine your mind following the stream of your breath through your body as it drops into the belly and eventually arises and comes back out.

Good.

[You might see some students giggling. Don't be surprised if this happens in the first three rounds of meditation! It goes away as students feel more comfortable and meditation becomes a normal part of math class.]

Inhaling, exhaling, your mind follows the stream of breath as it changes from its initial cool temperature beginning at the tip of the nose to a warmer flow on it its way out.

Continue doing this.

[Pause for 15 seconds – it's longer than you may think!]

Now take one final deep breath in, and let it all go through your mouth

[I make a loud exhale sound so students know what to do.]

Wiggle your fingers and your toes stretch your arms out and open your eyes.

There is a stillness and calm in the room. I myself also feel more composed since the initial class bell rang for period 3, six minutes ago.

And now the class's math adventure begins…

You may want to share some thoughts on the exercise with your students before moving on. While there is some conversation about the connection with learning and our minds within the class, I prefer the conversation to take place naturally and out of curiosity. I do make a point of giving closing comments as food for thought for the kids:

> *Take a note of how this exercise felt. Maybe it was very difficult to focus and it was a short but very uncomfortable journey to focus on the breath and sit quietly. It is with sitting with discomfort and practicing focusing that will also help us learn how to manage the chaos that often exists in our minds, an unfathomable immeasurable world, as we learn in school and go through the journey of our day outside of school and react to the people in our lives. Discomfort is not always a "bad" thing and often when we are doing math, things feel really uncomfortable but we have the power in ourselves to sit through it and learn!*

It's better not to force any conversation but rather allow students to soak in their first experience, as not all things can be intellectualized and analyzed. From my experience, students often ask to do the exercise again and, once I do it a few times, the students make a point to remind me if I forget to implement it in the beginning of class. The conversations generally flow organically as students also become curious about "all of this meditation stuff."

While I focused on the breath in this example, you may choose to focus on sound as strings of sounds from our environment and from within arise. Another focus point is touch – touch of clothing, the chair, and the air as it enters our nose.

After hundreds of mindfulness and reflective meditations with students in so many different settings, one more piece has also become very clear: *students notice that you care about them and they are more open and caring towards you as well.* It transforms the classroom atmosphere in intangible ways. There are times, as I discuss at the end of this chapter, when it transforms the very nature of how students interact with one another and learn mathematics. This began to take place more visibly when I ventured into regularly implementing reflective meditations. An important piece that should be emphasized is that students should be seated arm's length apart, or be allowed to find their own "meditation spot in the classroom" far from each other, so that they don't distract each other. Closing the shades on windows and lowering the lights also help create a calmer atmosphere in the classroom.

Overcoming barriers to learning math using reflective meditations

Reflective meditations allow us to converse and sow seeds through poetry rather than intellectual analysis or a how-to-lecture. Contrary to the unreasonable expectation that learning math is an effortless grasping of new skills or ways of thinking, learning with our full beings immersed in the sea of childlike curiosity and wonder is actually very difficult. Learning new math concepts and skills (or any subject matter) is not instantaneous and does not transpire without sustained effort. Being scared, lost, and upset is a normal part of learning. Not everything will feel easy and joyful.

To think learning is easy and completely comfortable is equivalent to thinking one can bypass the process of hiking arduous trails along a mountain to uncover the views from each step and still somehow have full-bodied experience of those breathtaking views at the top. Part of my goal is to normalize what is often seen as "abnormal." Normalizing the apparently "abnormal" is key to disruptive STEM education. The disruptive STEM paradigm seeks to render visible often invisible structures like systems of ability tracking that disempower both learners and teachers. Reflective mediation gives agency back to those closest to the learning process. What do students see as abnormal in their experiences as they learn math and consequently give up? Perhaps they view the prospect of practicing how to add fractions a dozen times without understanding as abnormal and give up. They perhaps see others learning faster than them and think that they must somehow be inferior or not good enough, not realizing that every human being has unique tendencies and ways of learning about the world. Reminding oneself constantly that material does not need to be understood in one day, or even one month, but may rather take sustained effort over months, is humbling but also empowering as it instills a faith in one's ability to navigate through a situation.

In other words, learning math is not meant to be a result-oriented pursuit from a given problem to an end solution. Mathematics is journeying through a puzzle, not knowing how and when the solution will crystallize. Unfortunately, this "find the answer"-based view of math, reinforced by so many environmental factors and internalized ways of thinking, takes a stronghold in young minds and needs to be broken open, unlearned, and replaced with correct notions of learning. Intensive standardized testing, of course, places an emphasis on finding the answer under timed conditions, and the process is often overlooked in the grading. In light of this, reflective meditations soften the impact of standardized testing on the mind and nurture curiosity and joy during the learning process. For teachers, and in my own experience, discovering the vastness and complexity of learning through practicing meditations is refreshing and an important reminder of how important the work of teaching is in an environment that so often undermines appreciation and funding for the work. In this way, teachers are integral to preventing bell hooks' mind–body split, which causes students to approach STEM learning in an excessively detached and purely intellectual way, as discussed in Gardner's introduction.

In addition to the provided script, here are three distinct scripts with background information and follow-up reminders that you can use with students or as inspiration to build your own scripts. The initial a mindfulness exercise is essential for setting the stage for these scripts that follow. It will be helpful to read through the scripts to gain both a deeper understanding of reflective meditations and a grasp on their diversity. The creativity of reflective meditations and the distinctive qualities of each one lie in their ability to resonate with the reader or listener who engages in them.

These scripts are not a one-size-fits-all approach but meant to be tailored to your students and can include your own variations and language. If you feel unsure about trying these exercises for the first time, you are not alone. In fact, that is part of the journey of learning! To break the ice with students, I first implement a brief mindfulness exercise (described earlier). There are a few important pieces to keep in mind before using these scripts:

1 If a student feels uncomfortable closing their eyes, do not force the student. Instead, give them the option of lowering their gaze to a still unmoving point in front of them.
2 Students dealing with certain psychiatric or other medical conditions may not be able to comfortably participate in certain exercises as easily. For example, a highly ruminative mind may get caught up with thoughts while focusing on internal anchors like the breath. Instead, visuals and stories with embedded messages may help this child. Script II, "Waves of the Mind," is an example demonstrating the use of a visual by comparing the mind with the ocean.

As seen in the first script earlier, the brackets include time-sensitive comments in the sense that they apply to the meditations at the moment you would be reading or practicing the script. They serve as guideposts for what can be done to make the practice more effective based on my experience.

Script II: "Waves of the Mind"

Background

In order to learn math, students must learn to sit with discomfort. This is especially true in an active learning format but also applicable in a more lecture-style format. It is not comfortable to crack open a problem or learn a concept that appears challenging at first sight, especially in front of peers. Frustration and nervousness at not being able to solve a problem or not understanding a lesson on the first round is normal. If all students in the class learn this message, they will feel more at ease being frustrated or being seen stuck on a problem.

Dealing with frustration and spending hours learning something new can be difficult for pre-teen or teen minds. The good news is that their minds are much more malleable than an adult mind, which has often become set in its ways.

Consistent frustration about being frustrated and understanding nervousness as a cue that an experience is impossible point to faulty preconceptions about the challenging and rigorous nature of learning. Some students feel sad or embarrassed if others are picking up material faster than them. There is some underlying wiring telling students that if they feel certain emotions or experience certain thoughts while learning, there is something wrong. It becomes essential to guide students on how to sit with their emotions and thoughts and normalize the presence of unpleasant experiences when learning math.

In order to train the mind to sit with the emotions that come up, I created the "Waves of Mind" meditation, where I use an ancient Indian metaphor of the ocean and its waves for the mind and combine it with the principles of mindfulness to help students develop a healthy distance from thoughts and emotions that bar the journey of learning.

Script

[Set up some background music that is soothing. I find it helps set a nice atmosphere for the students. Dim the lights a bit.]

> *If you are wearing glasses you may take them off and place them on your desk.*
>
> *Begin by sitting upright, drawing the shoulders back if they are slouching forward. I will also be doing the same.*
>
> *And close your eyes.*

[Students who are not able to keep their eyes closed may lower the gaze. Students that require the modification know by now after the first few series of mindfulness exercises, so they do not need to be reminded.]

> *Now, imagine your mind like a vast ocean.*
>
> *As you look at the ocean you see tiny ripples, and as you look closer and closer, you notice that these are waves of all sizes.*
>
> *Some large, some tiny, some faster than others.*
>
> *Give this ocean a color – perhaps deep blue – and picture the waves rising and falling into the vast ocean.*
>
> *The ocean stretches on forever and is unfathomable, for we have made this ocean the ocean of our minds.*
>
> *With innumerable waves rising and falling every moment, this is a dynamic ocean.*

There are parts that are more serene and parts that are more active.

Now, imagine diving into a tumultuous part of this ocean where waves are rising and falling.

As soon as one wave falls, the next wave picks you up.

You rise and fall.

You are constantly battling these waves and challenging these waves.

In no time, you are exhausted and you just want a break.

And so you walk out onto the shores of the ocean, perhaps on a beautiful sandy beach.

And you lie down, and watch the waves instead.

You realize that watching the waves rising and falling is so much more grounding than jumping into the ocean of waves.

So now, we have stepped out of our minds as we sit on the beach and we are looking at our minds.

And all these waves that rise and fall are the emotions and thoughts that run through us.

And often, emotions last for long, so the big waves represent our emotions and feelings.

The smaller tiny waves represent our thoughts, that arise and fall away more quickly.

And sometimes there are certain thoughts that don't leave us, so those may be medium to large-sized waves.

There is a lot of creativity in creating your ocean that represents your mind and waves that represent your thoughts and emotions.

Take a moment and observe what you feel right now and the activity of your mind.

If any thoughts arise, give the thought the form of a wave and imagine the wave picks up speed and grows tall.

We allow the wave to take its time to follow its journey, for all thoughts have a lifespan.

If we fight the wave, it lasts for longer as you splash around creating ripple effects and more waves.

So we don't challenge it – we just visualize it and watch it.

Before long the wave falls away and the thought extinguishes with it.

Draw and color your thoughts and emotions in as waves that rise and fall, remembering that waves of emotions will take longer to fall away.

And it's ok.

All waves are born and fall away.

For the next few moments practice doing this on your own. Anything that arises in your mind – sensations, images, thoughts, emotions – give them a form of a wave and just allow them to pass.

After all, you're the observer and you are not involved.

[Pause for 20 seconds – time it.]

Notice if you begin to engage with a thought and forget about the ocean.
Bring yourself back to lying down on our sandy beach and just watch.

[Pause.]

Good.
Wiggle your fingers and your toes, stretch your arms out, and open your eyes.

[As students are stretching…]

This is not easy and it takes a lot of practice. We all know that, when emotions arise we tend to react rashly and right away without thinking much. But with more and more practice, we can develop skills to deal with our minds and create a distance between us and our thoughts and emotions in a very healthy way. Then we have power over choosing to react in ways that are less hurtful to ourselves and others.

Reminders for the class

You can remind your students, as they work, that they will feel frustrated. You can precede a lesson by saying that feeling confused at first is normal and if something feels difficult, it is ok. These feelings will come and go. Let them know that you understand that, while you adapt lessons to the class, at times things will feel fast. If they feel someone else is getting a concept and they are still feeling lost, this is all a normal part of learning, but that does not mean that we stop working hard. We accept what is normal and continue trying and persevering. To sit with those emotions and keep trying and persevering will shape their long term mindsets much more than just giving up on a subject or topic because it is hard.

Script III. "Conversing with Struggle"

Background: I wrote this script in response to how adverse students, and humans, in general, are to struggle. I am not talking about suffering – I am speaking about struggle that will lead to being productive. There may be preconceptions that a certain amount of struggle is ok, but beyond that amount, it is better to give up and persevering is useless. It is entirely normal for students to try to avoid struggle, especially if they view struggle in a negative light. This script aims at helping students to build a positive relationship with and befriend a personified Struggle.

Script

> *If you are wearing glasses you may take them off and place them on your desk.*
> *Begin by sitting upright, drawing the shoulders back if they are slouching forward. I will also be doing the same.*
> *Take a deep breath in and let out a full exhale through the mouth*

["Haaaa..." make the sound for students to understand what you mean.]

> *As we close our mouths, you may want to move the lower jaw from side to side slightly until it finds a calm resting place.*
> *Allow the tongue to relax from the tip all the way back into the back of the mouth. This will help calm the chatter of the mind.*
> *On your next exhale allow the wrinkles of tension on your forehead to become smooth and relaxed.*
> *Notice if you are moving your hands or feet. Bring them to stillness.*
> *Allow your arms to become heavier on the exhale and stretch your hands out wide and then let them return to their normal position.*
> *Do the same with your legs on your next breath in.*
> *On an exhale, let the feet stretch their toes and return to resting posture.*
> *Allow the shoulders to relax and imagine tension in the neck dripping away like ice cream melting on a warm summer day.*
> *Now that you feel more at ease, you decide to have a conversion with an old friend named Struggle.*
> *And so you invite Struggle to sit with you.*
> *Before you do that, give Struggle a shape and living form. What does it look like and move like?*
> *It's a compassionate yet tough living, moving being.*
> *And you kindly invite it to sit down with you and converse.*
> *During this conversation practice breathing in the conversation and settling deep down into the essence and meaning of it.*
> *Now that Struggle is here in front of you, you ask Struggle with a capital "S":*
> *Why do you appear every now and then and make things so difficult?*
> *Struggle responds: It's not that you knock on my door and I appear from some external source.*
> *It's that I'm always present, in every moment, never leaving your side.*
> *You breathe in and you breathe out, absorbing all that you hear.*
> *And yet, you are baffled and so you ask Struggle: Why are you at my side and where are you in every moment?*
> *Struggle responds: I exist in the meeting of preconceptions and newly-learned information.*

When you absorb the experiences, facts, and observations of every moment, they meet with your prior beliefs and notions.

This meeting, a process of inquiry into what truly exists and which beliefs are true, gives rise to me – Struggle.

And for this reason, I never leave your side.

I am the phenomenon that, if met with wisdom, leads to clarity and evolution of the mind.

Again you breathe in and breathe out on this unique opportunity to speak with Struggle.

And you respond: And how do I resolve you if you exist in every moment, my friend?

Struggle responds: By leaving no question unturned, practicing perseverance, and seeking voices of wisdom.

By creating space for preconceptions and newly found information to mingle and all the while, accepting all the emotions that rise in this tumultuous process.

Wow! You think to yourself.

You take a deep inhale and exhale, thanking Struggle for the insights.

I have much to ponder, you say to yourself.

And you return to your breath, inhale, exhale.

And as you inhale, you breathe in what is left of Struggle's words and echoes.

As you exhale, you sit more deeply into its essence.

Breathe in, breathe out.

And when you're ready, wiggle your fingers and toes, stretch out your arms and open your eyes.

Remember to use a different tone for yourself and Struggle as you read aloud. Presentation of this script as a conversation is essential. You could try practicing this, or any of the other scripts beforehand. The more natural the implementation, the more effective the exercise will be at keeping the students engaged. You will also be able seep deeper into the ideas you are sharing with your students.

Reminders during class

Remind students that productive struggle is a normal part of learning and life and that it is essential for our growth. If students say that things are too hard, remind them that the feeling of something being too hard is just a message to keep digging deeper, and drop hints and give some guidance on the math problem. If students are working hard on a problem, compliment them for sticking by Struggle's side.

Script IV: "The Magical Box"

Background

As in my anecdote about the first mindfulness exercise of the year, students bring high energy and plethora of experiences and backgrounds to class. Each student is a universe, and unless learners create space in their minds to learn, the mind will forever be tugged by thoughts about the past, the future, gossip from the hallways, home lives, body image, upcoming exams, what to eat, and so much more. This is quite overwhelming. For a young mind, each lesson and every new learning experience is a journey. And just as we all prepare and pack before journeys, it is important for students to have this opportunity to prepare for their learning journey as well. This script helps students let go of distracting thoughts, feelings, and emotions as they place them into a safe magical box.

Script

Go ahead, you may take off your glasses and put them on the side.
Get into a comfortable position and close your eyes.

[Students who are not able to keep their eyes closed may lower the gaze. Students that require the modification know by now and you don't need to repeat it every time.]

Before any venture into learning anything new, it's important for us to bring our entire selves to the journey of learning.
And so, take a deep breath in and out.
Feel free to stretch your mouth open and close it to relax the jaws.
Raise your eyebrows and drop them, relaxing the wrinkles of worries from your foreheads.
Allow the eyes to sit deep into the sockets as the eyelids softly cover them.
Notice if you are feeling fidgety or if there are a string of thoughts circling through your mind. Before we begin class today, we are going to tuck away all possible distractions so that we can be fully present on our learning journey.
So imagine a special tiny box hovering in mid-air right next to you.
It locks itself and opens itself just to you and nobody else in the universe.
Notice if there are things on your mind that are bogging you down.
Imagine these floating out of your head and into the box which has now opened itself for you.

Notice if there's tension in your body or you feel tight in your jaw, your shoulders, or your back. On an inhale collect all this tension and on an exhale send it right into the box and out of your body.

Notice if there are any thoughts in your mind that say, I don't want to learn today.

Allow these unhelpful thoughts to also float out of your head and into the box.

Notice if there are any thoughts that say, I'm not good enough or I am not smart enough to learn. These are thoughts that are trying to distract you from the truth that you are whole, capable, and 100% OK.

Allow these thoughts to float out of your head and into the box.

Take a deep breath in and a full breath out.

One by one, take your time, and allow anything that will get in the way of your ability to learn today to flow out of you and into the box.

[Pause for 20 seconds – time it.]

Now it's time for the box to close itself so send any remaining thoughts, feelings, or habits that you know will hamper your ability to learn.

[Pause for 10 seconds – time it.]

Now the box magically locks itself and floats down becoming tiny tiny tiny, almost like a dot, and enters your pocket.

Nothing in this box can be touched by anyone but you and will be available to you to reaccess once class is over.

But for now, with a full heart for diving into a new adventure, we embark on a journey to learn in class today.

So go ahead, wiggle your fingers and your toes, and stretch out your arms. Open your eyes. Good.

Reminders for the class

Remind students that if distractions come up, they can still use the magical box and send distractions into it and bring their attention back to the task at hand. Students may randomly begin to do silly things in class at times (as all teachers know!). A little silliness helps learning as it relaxes the class but too much is a distraction. This is the time to maybe just go over and whisper to them to use the box.

Reflections on experiences

After implementing these exercises, you may be surprised to find students repeating pieces of the meditations. Sometimes, they even advise what a peer should do if they say, "I give up." They will respond with a phrase from one of the meditations I have done with them. One of my students told me that he used my exercises before basketball games when he was nervous. Students pick up on the messages as they sit absorbed in the script and music, with eyes closed, and the messages trickle into actions during class and outside of class. I have consistently seen students improve in their collaboration during group work as they keep asking questions and I drop hints. While not included above, after a meditation on embracing one's pace in life, students who were anxious high achieving students began to take time to slow down and help each other more. I have noticed, again and again, that students also begin to see you as a more caring figure.

In my experience, very few students have shunned the idea of joining the class for these exercises. On the other hand, they embrace the activity. For some of my classes, it became a ritual. The students would be sure to turn off the lights and prepare their desks to begin meditation before lessons. I have noticed students treating each other more kindly and taking more time to teach concepts to one another. They have also expressed how it melts away a lot of the stress they go through and refreshes the mind. Students who have a tendency to disrupt class are more focused and cooperative. Every class, I have a few students that feel a bit sleepy, and I remind the students that the body is a great messenger – we must take care of ourselves and rest.

Teachers love collecting observations to better understand our learners' needs. Teachers are sensitive to changes in the environment and the way students are treating others or going about tasks; we love watching journeys of growth and change. Often, it is useful to ask students to anonymously jot down some thoughts after the first or a few rounds of exercises. This way, you get a direct response from the students. You can give them a quick Google Form for them to fill out or you may just have students respond to a question on a piece of paper.

Following the events of the 2020 pandemic and an increased use of technology in education, using meditation also provides educators with an opportunity to connect with students on an online platform. While you may be miles apart and not in the classroom together and feel sad at the lost opportunities to connect with students, these exercises have immense potential. They let students know, no matter how far apart the class may be, that they share some very profound experiences as a group. Our minds are forever changing, we all struggle, we all try to persevere, we all get bogged down by thoughts and emotions, and we all need guidance on what learning is and what it takes to learn. From my experience in 2020, meditation has been very helpful in connecting with students and helping them traverse the tricky terrain of learning that can arouse a vast spectrum of emotions.

Reflective meditation as disruptive STEM teaching

The very nature of carrying out reflective meditation as a concept and ongoing practice is a part of disruptive [to current] STEM education. Making individuals aware of the mind–body connection and its importance in enhancing their ability to process and be available to learning in each moment is not what the typical student experiences when walking into their STEM classroom. Here, criticality meets subversion. Practicing reflective meditation counteracts the idea that only the intellect dances to create and cultivate advances in STEM and STEM education. The very nature of implementing meditation is an antidote to the mind–body split hooks (1994) speaks of.

Today, learning is tested by the end result. At every turn, students are bombarded with high-stakes testing. However, the very nature of learning, as I argue earlier, is process-oriented. The very act of turning inward and guiding students to do so as well is an act of love and an initiation into the practice of self-acceptance and non-judgment from the teacher. This happens in the crucial transitions of stepping from known to unknown territory in the mathematics classroom. Stepping into the unknown (see scripts "Conversing with Struggle" and "The Magical Box") is handled with sensitivity and care, an arena where students often only categorize themselves to be "successful" or "failing."

Reflective meditation scripts introduce the student to the mind and its qualities, involving the whole self. Often students enter the classroom with what we could call the "math self," judged on a spectrum from "good" to "bad." Disrupting the notion of "math self" and inspiring the notion of "whole self" blur the delineated parts of the brain. Instead, the child experiences one whole moving, processing mind immersed in a journey of learning.

The interdisciplinarity of the concepts put forth in this chapter spans and is shaped by subjects and philosophies developed over millennia. As emotions and the process of creating and playing with ideas fuses with the analytical and result-driven tendencies of the mind, STEM and non-STEM studies fuse. It is as if we touch the humanities, rich in emotion, and social sciences, rich in psychology, within the otherwise abstract math classroom.

These philosophical and contemplative perspectives – many of which are drawn from pre-modern India – transform the preparation of the mind to dive into the world as a whole, which includes STEM and non-STEM explorations. Instead of wandering into the wilderness of the journey of learning without direction, the student now forms an active network of anchors that they can use to navigate their internal and external worlds. By the end of this process, we have encompassed the entire student.

References

Delpit, L. D. (2012). *Multiplication is for white people: Raising expectations for other people's children*. The New Press.
Hooks, B. (1994). *Teaching to transgress*. Routledge.

10

STEM TEACHING FOR COLLECTIVE CARE AND ACTION

Margery Gardner and Hugh Burnam

Introduction

> My foundation is Freire and hooks and [the] concept of teachers and learners working together and everyone being a learner and creating knowledge together. Also, understanding that as a teacher you should never do things that you don't want to do yourself. You need to care for yourself in order to care for your students. The warm demanders piece is important especially since I came from a trauma background. It's important to hold [learners] to high expectations … Ladson-Billings' Multiculturalism is important because you can build off of knowledge they [students] already possess. We also need to give the tools to succeed given discourses of power-Lareau, Delpit. I want to be involved in activism. Coming together with students and the community to transform problems we all see.
>
> *(Interview, 12/16/2020)*

Maria, a teacher preparation program completer in the Northeast, outlines her philosophies and connections to theory in this passage. She demonstrates a disruptive teaching perspective through the inclusion of scholars and ideas drawing from critical, subversive, interdisciplinary, and identity-forward lenses. We opted to include Maria's narrative to share how one teacher makes sense of disruptive concepts and forges generative bonds between theory and future practice. Maria articulates the evolving process of identity development, listening, and unlearning that is essential to the

DOI: 10.4324/9781003395782-14

critical work of teachers. As hooks declares, we also encourage our readers to view theory as a site for continued reflection and healing. This passage reveals many of the internal hopes and desires that we have for disruptive STEM education as a movement for change.

Maria provides insights on the future of disruptive teaching that continuously engages theory.

As we unpack Maria's narrative, we notice her attention to self-awareness and personal connection to students. By sharing her trauma background, Maria models the vulnerability that actualizes her words. One of the many critical education scholars she names, Lisa Delpit, advocates for an environment that provides support as well as academic challenge, especially for youth with trauma backgrounds as Maria suggests. As discussed in Chapter 1, Delpit (2012) describes the concept of "warm demander" as grounded in a multidirectional relationship that presses the learner to build academic capacities through a humane and loving treatment. Pedagogically, Maria's philosophies embrace play-based, student-centric learning approaches. Maria, now a middle school teacher, similarly recognizes the human need for connection and care that can be fostered through teacher attentiveness and affirmation. Maria displayed reflective and reflexive characteristics that can support her critical teaching work. Maria's heavy emphasis on critical scholarship from folks such as Freire, hooks, and Ladson-Billings exemplify a strong commitment to understand how power dynamics function within schooling and society.

Revisiting major themes

Disrupting Secondary STEM Education: Educator Experiences of Teaching for Globally Just Futures sheds light on the intellectual, emotional, and spiritual work of teachers as they make daily decisions within particular school and world contexts. Authors share interwoven narratives of STEM teachers from dynamic and intersectional identities that open the black box to share how internal choices are made. The work of teachers represented across a career spectrum from teacher candidates to teacher educators offers perspectives from different professional vantage points. This includes David's story in Chapter 3, a student teacher who brought to the surface how US political intervention in Latin America shapes discourses around immigration today. Providence and Enrique, authors of Chapters 6 and 7, are relatively new to teaching and bring criticality through queer perspectives and pedagogies that expand student capacity to understand how social forces map onto math and science. Educators like Anita and Randa, co-authors of Chapter 8, continue to grow through professional collaboration that transcends the boundedness of their own affiliations.

Thick descriptions of STEM disruptors in Chapter 1 illuminated histories and place-based complexities that add texture to conversations around subversion, solidarity, and humility. Jason actively worked to disavow the narrative that social justice math teaching meant being a "nice" White teacher. Jeri confronted overtly racist acts in order to celebrate Black culture and bring critical literature into her biology classroom. D. transcended stereotypical understandings of what Blackness looks like in predominantly White, rural school spaces. Shanaya empowered her math students to understand the subject through an ethnic studies framework. Alongside these portraits of disruptive STEM teachers are the stories of the book's authors, who describe their own journeys through introspection and sharing of curricular artifacts that illustrate the pedagogical realities of their critical work, all seeking answers to "questions worth knowing." Like Maria, these contributors seek to grow and learn from their surroundings and bring generative experiences to their classrooms.

This book also recognizes that the labor of radical re-imagination is often distributed unfairly to place burdens on teachers whose identities are under constant attack by White supremacist hegemony. The de-centering of Whiteness and normativity is hugely important to the viability of disruptive STEM teaching. This book purposefully leans on an array of educators that identify in complex and intersectional ways. These critical identities send a collective signal to our students that we all work in solidarity to do better. The personal identity dynamics within the authorship of this work are celebrated and used to ground commentary and analysis in reflective ways.

We all seek answers to "questions worth knowing." This work extends what is already known about the critical STEM literature by assembling conversation around the secondary level where access can be harmed through practices such as tracking. The primary objective of this book has been to disrupt the common tropes associated with STEM education and replace them with a reimagined future that decenters Whiteness and White supremacy. "A goal for critical science education scholars should be to render science education discourses in such a way that they can no longer stand the way they stood before" (Bazzul, 2015, p. 423). Teaching is a complex endeavor that is subject to social pressures, capitalistic tendencies, and structural racism, to name just a few challenges. The book took readers on a deep dive into the work of teaching at the secondary level and beyond in order to dislodge preconceived ideas about STEM education as objective, unbiased, and more valuable than arts and humanities. We investigate how specific teachers use curriculum as a tool to speak to students about broad social issues such as initiatives to defund the police or the commercial fishing industry. Rooted in tangible tools, including curricular examples, we show specific ways that teachers navigate structural terrains and context-specific challenges in order to speak their STEM education truths. In the next section, we offer commentary and perspective on the pressing needs of Indigenous students and how our efforts as disruptive STEM educators necessitate both hope and love that can improve educational access across an array of communities.

Supporting Indigenous students in STEM

Early in our writing process for this chapter, we talked about the problem of Indigenous erasure in schools. They reflected on ways that people often recall how they "learned" about Indigenous peoples when they "built a Longhouse" sometime during their primary education, and that was it. For years of our P–12 education, the only knowledge that most people have about Indigenous peoples is recalling the time they built a replica of a Longhouse.

We assert there is a critical importance for educators to use a culturally responsive and sustaining lens to address gaps in school curricula around Indigenous education and to support Indigenous students. Critical race scholar, Aja Martinez (2020) writes:

> The core curriculum is the occupied space of white racialized perspectives, while the voices and stories of the racialized 'other' are pushed to the margins in elective courses, at best, or footnotes and asides within the core curriculum, at worst (p. 117).

Educators need to recognize that as long as Native peoples are the optional class, taught as if solely in the past, breezed over, or reduced to simply – "the time we built a longhouse" – their minimal efforts reflect a long history of Indigenous erasure.

It is also not enough to recognize Native peoples simply as a racialized group; instead, schools should acknowledge the "special relationship" that Indigenous nations have with the United States government in a nation-to-nation relationship, specifically recognizing the important role that Indigenous sovereignty plays, not just in the community, but in education too. Schools must also implement policy to publicly and regularly recognize the Indigenous lands in which their schools occupy. To do this accurately, superficial acknowledgment for public performance is not enough, rather, this work should be done in collaboration with local Indigenous community members to consider specifics and context to the land and should be a long-term partnership (Robinson, Hill, Ruffo, Couture, & Ravensbergen, 2019).

In order to accurately represent Native peoples in the school curricula, educators should work diligently to uphold long-standing relationships with local Indigenous nations, tribes, and community members. Indigenous peoples must be given the opportunity to take the lead in representing themselves however they see fit in their local schools and communities. Educators need to continue to explore ways in which a "multicultural" education actually erases Indigenous civic engagement and the meaning of Indigenous sovereignty. Sabzalian (2019) writes of the importance of upholding Indigenous sovereignty in education, "cultivating meaningful partnerships with local Indigenous peoples, organizations, or nations is not only a policy imperative, but a foundational relational practice if public schools are to harness the power of citizenship education to uphold and support tribal sovereignty" (p. 333). According to the National Indian Education Association (NIEA), it is up to each state to provide unique opportunities to "promote tribal sovereignty in education and create schools with positive and inclusive learning environments for Native students" (National Indian Education Association, 2023).

Racism and settler colonialism are systems which reflect a long history of oppression for enslaved Africans and Indigenous peoples. In one such example, the US government forced Indigenous peoples to assimilate to US society through residential boarding schools, stole Indigenous land through fraudulent treaty-making, and attempted to eradicate Indigenous culture and languages through ethically immoral practices often acknowledged as attempted cultural genocide. Teachers must continue to address this history in their classrooms and explore ways in which this system still operates today (and not just in the past). In their study about misrepresentations in US history standards in P–12 education, Shear, Knowles, Soden, and Castro (2015) found that education "standards in the United States largely depicted Indigenous peoples as existing in the distant past and are thereby marginalized from the American present. First, Indigenous peoples are cast as outsiders to the master narrative" (p. 83). Teachers have the opportunity to create spaces for students and colleagues to discuss ways that systemic racism and settler colonialism still impact Indigenous peoples and so many communities to this day.

While it may seem challenging, teachers should continue to build self-awareness about their own positionality and set of privileges that they bring to the classroom, whether through race, class, nationality, gender, sexuality, ability, religion, or more, and recognize their positionality and privileges as inherited from systems of inequity. This sort of reflective work is demonstrated through the words of Maria at the start of our discussion. Teachers are not the saviors of Indigenous students and peoples, but instead can use their privileges in allied advocacy to provide a platform for Indigenous communities to empower themselves (Sabzalian, Shear, & Snyder, 2021). Teachers are encouraged to unpack their own perceptions (or misperceptions) about

Indigenous peoples – which, due to invisibility of Indigenous peoples in the curriculum, in media, and public discourse, may be informed by stereotype and misrepresentation. To address this issue, teachers should create spaces for Indigenous community members to share their stories, their work, their art, and their own histories of the land, creating a co-learning environment, not just for students, but for teachers and school community alike. All students should experience the land outside of the confines of the classroom and begin to explore the natural world around them and continue to cultivate their relationship to the land by bridging Indigenous worldviews with STEM education. Developing counternarratives that can resist normative views of STEM can uplift voices frequently silenced in schools.

In an in-depth literature review for supporting Indigenous students in STEM education, Jin (2021) indicated that out of 24 studies about Indigenous students in STEM, the outcomes were as follows:

> 21 (88%) measured and reported outcomes related to Indigenous students' relationship with science, including science learning, science-related careers, and attitudes toward science. 11 studies (46%) presented outcomes related to Indigenous students' relationship with their own cultures and traditional knowledge, including understandings and pride of Indigenous knowledge and traditions. 6 studies (25%) reported outcomes around Indigenous students' awareness and perceptions of the connections between Western science and Indigenous cultures.
>
> *(p. 9)*

Jin reiterated the importance of designing both formal and informal education contexts for Indigenous students. For many students, science can be personally relevant, and informal contexts (such as summer programs and after-school programs) may help Indigenous students connect Indigenous knowledge with Western science. STEM education in P–12 settings is plagued by the underrepresentation of Indigenous teachers and boasts very limited support for non-Indigenous teachers. This is problematic since schoolteachers have such a vital impact on Indigenous student learning. The teaching profession at the secondary level remains predominantly White. Jin writes that in order to best prepare Indigenous students, teachers must continue to value the cultural and experiential knowledge of their students, that "instead of putting Indigenous knowledge and Western science knowledge in opposition to one another, these culturally relevant programs extend knowledge systems and find value and new perspectives for teaching and learning from both" (p. 12). Lastly, three broad outcomes for student development are relevant to diversity in STEM and/or Indigenous resurgence in STEM, "(1) outcomes related to Western science or STEM;

(2) outcomes related to Indigenous knowledge and culture; and (3) outcomes related to the complementarity between Western science and Indigenous science." (p. 12). As students move across cultures between home and their Western education, educators can help them by making STEM more relevant to Indigenous students and thus attract more Indigenous students to STEM fields, which would in turn make STEM more diverse and welcoming to students from underrepresented populations disrupting the notion of who can "do" science.

For Indigenous students, the practice of maintaining a caring and respectful relationship with the land is more than a goal; it is a cultural imperative. Traditional knowledge of Indigenous communities suggests that people are not separate from nature but rather an integral part of it. Therefore, Indigenous knowledge teaches the critically important responsibility of Indigenous peoples to protect the land. By honoring their sacred relationship to the land, Indigenous communities know that they merely borrow its use from the coming generations during their short lifetimes. Thus, their cultural imperative is to care for the land so that future generations may also live in the same reciprocal balance with the natural world that they were once gifted.

Our collective future depends on bridging STEM research and traditional knowledge from Indigenous, Black, and Brown communities – coincidentally our most vulnerable populations during environmental catastrophe. Prior to his passing, scholar John Mohawk (Seneca) once spoke at a conference and said:

> All of the survival techniques we learned about our relations to cultivars and everything at this hour stands in peril. And our relationship to wild plants stands in peril. The big human relationship to our cultural heritage is on the verge of extinction, and we need to change that.
>
> *(Mohawk, Bioneers Conference, 2004)*

Those Indigenous generations are here now. Indigenous youth are students in P–12 schools, students who bring this worldview to their education. As educators continue to welcome Indigenous students and communities into their spaces, it is imperative that Indigenous students learn scientific approaches to STEM education and research. They will undoubtedly bring this knowledge back to their communities to help maintain their Indigenous nations. But likewise, STEM educators and entire fields of study, learn from Indigenous students and other communities of color through the incorporation of ancestral knowledge into their research and innovation.

Disruptive STEM does not need to be a contentious space; in fact to so many populations whose perspectives often go overlooked, it is a space of unrelenting hope. As humans, we all too frequently embrace systems of our own creation, systems of settler colonialism, patriarchy, and capitalism, all human-made systems which have perpetuated inequities for hundreds, if not, thousands of years. We are currently facing the consequences of our systems, which caused us to overlook our relationship to the lands in which we occupy. Globally, we stand at the precipice of what is being called "the Anthropocene," a geologic epoch of catastrophic environmental global change caused by large-scale human activity. Human beings are not living in reciprocity with the lands in which we live; so many humans have lost connection to that which gives us life. But it is not too late. Educators, administrators, policymakers, and community members have important responsibilities to uphold, given our current geologic epoch – they can create opportunities for diverse groups of students to thrive in STEM fields to address our current realities.

While we are solely responsible for the systems that we create, we also have an urgent and critical responsibility to create new systems which may look back, using traditional ancestral knowledge, to those who lived in reciprocity with the earth. As we learn from the past, and we look forward, may we think of the coming generations in our minds, and may their well-being be our cultural imperative.

Disruption as hope

In order to move forward, we recognize that political work is involved in the process of change and hope as central to the work of disruptive STEM teaching. Schools who seek to dismantle faulty systems are under attack politically now more than ever in recent history as people in positions of power attempt to uphold White supremacist systems and structures. To sustain their employment and financial stability, disruptive STEM teachers often move this work underground, where it can be protected using subversive acts. This book is an act of resistance and solidarity with disruptive STEM teachers that affirms this work despite pressures that threaten to silence it. By sharing experiences from other STEM educators, it is our intention to energize a committed base with a theoretically grounded common story. We assert that disruptive STEM teachers possess the necessary social and resistant capital to enact broad-based change (Yosso, 2016). Joining forces with the community, in particular, student family members, can support educational change as a fully public endeavor and an intersectional movement for justice.

Traditions of organizing around issues of social justice hold lessons for educational change. Oakes and Lipton (2002) consider educational change as a social movement that is composed mainly of three features: relationships, generative dialogue, and action. Through the creation of interactive networks, now primarily accomplished through social media outlets, individuals amplify their voice and enhance the collective social capital necessary to evoke change. The group shares responsibilities and distributes internal leadership to ensure that decision making aligns with the desires of the whole. Generative dialogue in the tradition of Freire (1970) allows for collective identity development based on a unified story of who the group is, what they do, and why they do it. Organizers identify incremental objectives that seek to disrupt existing power structures in order to press further. Resources are funneled to meet objectives in a strategic and timely manner. "Such focused campaigns do more than advance the cause that the community has identified; they also develop the community itself" (p. 399). However, we note Oakes and Lipton's (2002) caution that change does not come without trauma and achievement is non-linear. Combating White supremacist sensibilities is an intensive endeavor that involves much emotional labor and often comes at professional risk.

We need to look carefully at the contours of educational reform that shaped and continues to shape collective experience especially in the wake of COVID-19 that has shaken our educational systems in traumatic ways. A normalized survival mode mentality neglects learners as whole people in need of healing and care. Rhetoric surrounding academic losses as a result of the global pandemic overwhelms our sensibilities as a school collective for what is ultimately best for teachers and learners. In the rural schools where we ground much of our work, fears of below average performances in ELA and mathematics causes administrations to lengthen instructional periods for these subjects and focus on base level skills such as grammar and fluency. Science and social studies are considered boutique subjects, only afforded time and energy in the unlikely event that "mastery" of ELA and mathematics happens to occur. These decisions work in opposition to what we know as educators is pedagogically best. The bureaucratic demands of the system fall unequally on certain educators who must then lean more on subversive acts to satisfy and sustain their classrooms. To enact disruptive STEM under these conditions requires thoughtful community building among teachers, movement building in coalition with broader communities, and audacious hope.

Locating hope in our work as disruptive STEM teachers involves deep introspection. In Duncan-Andrade's (2009) poignant work on hope in urban schools, he sets forth three pathways to understand trauma and how educators can lead students toward finding ways to heal and empathize with others. In alignment with Cornel West's version of critical hope, Duncan-Andrade brings into focus material, Socratic, and audacious elements. Material hope urges educators to push beyond the parameters of standard-based outcomes as narrow measures of achievement and to think about critical pedagogy and justice-focused topics as fully embedded in instruction. For instance, Jeri expends her own financial resources to purchase books for her students as part of her classroom library. The teacher in some ways also acts as a material resource that students can draw on to enhance their ability to interpret complex events and to make sense of their world and advocate for re-imagination.

Hugh brings into focus the harms of the settler colonial project to silence Indigenous people. Socratic hope requires both teachers and students to painfully examine our lives and actions within an unjust society and to share the sensibility that pain may pave the path to justice. This form of hope affirms an emotional response by youth, especially BIPOC youth, of anger, pain, and resentment. It acknowledges these feelings as real and part of a process of healing and being human. This approach treats learners as extended family members of the teacher. Learners are held to a high standard, as if they were teachers' own children. The concept of warm demanders that expects achievement though a paradigm of care connects closely with Socratic hope as a concept. Audacious hope calls for a radical healing of systems to better function in service of the children and families. Learners are deemed endemic to the classroom and their removal or isolation is an impossibility. Education centers on the whole child and prioritizes their holistic well-being. Hope becomes dimensional in this way, viewed as a full investment worthy of challenge, sacrifice, and pain. Superficial calls for diversity, equity, and inclusion in schools represents a hokey version of hope that reveals a lack of true motivation for legitimate change. Embracing audacious hope during a time where global environmental and health crises rest heavily on us all is the way forward (Duncan-Andrade, 2009).

Disruption as love

The work of disruptive STEM teaching includes savvy navigation of school contexts guided by a loving orientation. There are an ever-changing set of actors and political priorities that teachers must be aware of. Freire (1968/1970) refers to this tension as teachers straddling two worlds, those of the oppressor and oppressed simultaneously. These reflective and reflexive stances are ever present in the evolving work of teaching. This book explored how teachers think about their practice and broader positionality within school systems. The work of disruptive teaching described in this work as both fully intellectual but also spiritual and embodied. It is an act of love.

Critique of systems and practices is a clear demonstration of love. Reyes, Radina, and Aronson (2018) explain love as an act of resistance that consists of vulnerability, collective support and healing, and critique. "When we step into the classroom we must bring the messiness of who we are with us" (p. 826). Vulnerability to expose our full identities, including myself as a cis-gendered female, White teacher, generationally complicit in the systems of today, allows for damaging histories to surface. It acts as a starting point for change and as seeds for Camille Dungy's (2020) garden. This element is something that Maria picked up on in her statement at the opening of the chapter, *as a teacher you should never do things that you don't want to do yourself. You need to care for yourself in order to care for your students.* Through the vulnerability of sharing our stories, collective support and healing can be realized. While sometimes painful, critique strengthens growth and allows it to be directed in productive and targeted ways. Critical expressions must center on teacher preparation, questions of power, and continual growth through introspection. The space to enact loving resistance can be created by promoting environments conducive to dialogue and counter storytelling.

References

Bazzul, J. (2015). The sociopolitical importance of genetic, phenomenological approaches to science teaching and learning. *Cultural Studies of Science Education, 10*, 495–503. https://doi.org/10.1007/s11422-014-9605-0

Delpit, L. D. (2012). *" Multiplication is for white people": Raising expectations for other people's children.* The New Press.

Duncan-Andrade, J. (2009). Note to educators: Hope required when growing roses in concrete. *Harvard Educational Review, 79*(2), 181–194. https://doi.org/10.17763/haer.79.2.nu3436017730384w

Dungy, C. T. (2020). Reasons for gardens. *Ecotone, 16*(1), 154–158. https://doi.org/10.1353/ect.2020.0029

Freire, P. (1970). *Pedagogy of the oppressed* (M. B. Ramos, Trans.). Herder & Herder. (Original work published 1968).

Jin, Q. (2021). Supporting indigenous students in science and STEM education: A systematic review. *Education in Science, 11*(9), 555. https://doi.org/10.3390/educsci11090555

Martinez, A. Y. C. (2020). *The rhetoric and writing of critical race theory.* National Council of Teachers of English.

Mohawk, J. (2004). Survive and thrive. *Bioneers National Conference. Indigenous Knowledge*, Vol. 1. Retrieved: https://bioneers.org/john-mohawk-survive-and-thrive-bioneers/

National Indian Education Association. (2023). State priorities and resources. https://www.niea.org/state

Oakes, J., & Lipton, M. (2002). Struggling for educational equity in diverse communities: School reform as social movement. *Journal of Educational Change, 3*(3), 383–406. https://doi.org/10.1023/A:1021225728762

Reyes, G., Radina, R., & Aronson, B. A. (2018). Teaching against the grain as an act of love: Disrupting white Eurocentric masculinist frameworks within teacher education. *The Urban Review, 50*(5), 818–835. https://doi.org/10.1007/s11256-018-0474-9

Robinson, D., Hill, K. J. C., Ruffo, A. G., Couture, S., & Ravensbergen, L. C. (2019). Rethinking the practice and performance of Indigenous land acknowledgement. *Canadian Theatre Review, 177*(1), 20–30. https://doi.org/10.3138/ctr.177.004

Sabzalian, L. (2019). The tensions between Indigenous sovereignty and multicultural citizenship education: Toward an anticolonial approach to civic education. *Theory & Research in Social Education, 47*(3), 311–346. https://doi.org/10.1080/00933104.2019.1639572

Sabzalian, L., Shear, S. B., & Snyder, J. (2021). Standardizing Indigenous erasure: A TribalCrit and QuantCrit analysis of K–12 US civics and government standards. *Theory & Research in Social Education, 49*(3), 321–359. https://doi.org/10.1080/00933104.2021.1922322

Shear, S. B., Knowles, R. T., Soden, G. J., & Castro, A. J. (2015). Manifesting destiny: Re/presentations of indigenous peoples in K–12 US history standards. *Theory & Research in Social Education, 43*(1), 68–101. https://doi.org/10.1080/00933104.2014.999849

Yosso, T. J. (2016). Whose culture has capital?: A critical race theory discussion of community cultural wealth. In *Critical race theory in education* (pp. 113–136). Routledge. https://doi.org/10.1080/1361332052000341006

APPENDIX

Chapter 8. Energy Return on Investment

1	1,235	355	11,123	27,332
- 1	X 16	+ 15	X 357	X 567

25	1	10	34,001	31,000	40
+ 75	+ 2	X 55	X 995	- 150	X 9

4	3,984	5	99,999	72,911	100
+ 3	X 44	+ 6	X 888	X712	X 100

20	888	67,823	12,994	3	2,345
- 15	X 5	X 194	X 715	+ 12	X 11

50	56,234	77,000	34,267	668	7
+ 35	X 823	X 5	X 519	X 832	X 9

1	33	34,956	7,936	25	573
X 12	X 5	X 327	X 83	X 11	X 111

56,067	33	77,992	666	94,825	67,206
X 376	X 5	X 810	-123	X 789	X 739

Chapter 7

Appendix A: The "City Budget Project" Assignment Description Provided to Learners Using the GRASPS Model

Goal: In an effort to apply a social justice lens in math, the goal of this task is to have you identify and evaluate the (non)proportional relationship of the San Antonio city budget over the past three years, paying special attention to the increase or decrease of funding for public institutions (libraries, parks, hospitals, public safety, etc.). You will graph the rate of change for funding of at least three public institutions and identify whether the relationships are proportional with one another: Is funding for San Antonio public institutions increasing or decreasing at a constant rate of change?

Role: Citizen of San Antonio, a tax payer

Audience: You need to convince city officials: Mayor Ron Nirenburg, City Manager Erik Walsh and your Council(wo)men as to why funding should be reallocated across three San Antonio public institutions of your choice.

Situation: You are a curious citizen of San Antonio wondering how your tax dollars are being used to fund public institutions across the city. You find the approved city budget of San Antonio for the past three years listing the amount of money going into public institutions, such as hospitals, libraries, parks, and public safety.

As you look through the city budgets for the past three years, you notice changes in funding for several public institutions; funding is either increasing or decreasing for several public institutions. You are interested to see if funding is increasing or decreasing at a constant rate of change for three public institutions. You want to ensure that the city is dispersing your tax dollars across all public institutions equally, and that no institution is increasing or decreasing at a varying rate of change that takes money from one public institution to fund another.

You will do this by comparing the rate of change for public institutions within the past three years and identify whether that rate of change is proportional. You will choose three public institutions, focusing on institutions that have both increased and decreased over time. Once you have gathered this information, you will represent it in a linear graph showing the increase or decrease in funding of each public institution over the last three years. Once you have gathered this information, you will write a letter to city officials expressing how you believe your tax dollars should be spent using your findings as evidence.

Product: You will write a letter addressed to the San Antonio Mayor, City Manager, and your Council(wo)man containing the following:

- An opening statement introducing yourself
- A claim on how funding should be allocated for the upcoming city budget, and why
- Evidence that supports your claim. Communicate the work you did that supports your claim:

  - Provide a written comparison of the relationships between the funding of three public institutions over the past three years using mathematical vocabulary and the order laying out your computations

- Provide three linear graphs as a visual representing the (non)linear relationship of the increase or decrease in funding for each public institution (y) over the past three years (x)
- A reasoning statement where you are showing city officials how the evidence presented supports your claim on budget allocations
  - Sentence starters to help are

    - This supports my claim because...
    - These visuals suggest...
    - If ... then...

- A closing statement restating how your claim will better the greater San Antonio community

Standards: A successful outcome will be a strong understanding of how proportional relationships exist outside of the classroom via the letter you write to city officials. I will be assessing your ability to accurately identify the (non)proportional relationships of funding for three San Antonio public institutions over the past three years. I expect to see your computations.

Formatting: Your letter should be a Google document, double-spaced with a word count of 300–500 words. Your letter will follow the CER formatting used in ELA that answers the Research Question aforementioned.

Appendix B: Learner Work Sample, Learner Provided with the Pseudonym of "Taylor Gomez"

To the office of Mayor Ron Nirenberg,

My name is Taylor Gomez, and I am currently a 7th grader attending the Emma Tenayuca School for Women Leaders located here in San Antonio. This year, we have learned quite a bit, and two of those things were proportional relationships and budgets. At this time, my peers and I have been presented with a project to choose at least two money allocations and decide whether they are something that we, as future taxpayers, agree or disagree with and why. The earmarks that I have chosen are Health, Animal Care, and Government and Public Affairs. I have chosen these departments because I believe that the health of your people should be top of your priority, as well as their pets, along with the community that the people live in. For the annual budgets yet to come, I would assume that the people would best benefit from a slight increase in Government and Public Affairs, Health, and Animal Care. This is because our community is slowly falling into unhealthier eating habits resulting in obesity rates increasing, which leads to the need for places of exercise that are accessible to everyone, like parks.

Animals are lifelong companions to many, veterans tend to have service animals but cannot afford to properly take care of them.

While comparing the allocations from 2019 to 2021, there are visible increases in Health but not as much in Government and Public Affairs and Animal Care. As you can see in the linear graph, red represents Health, green for Animal Care, and blue for Government and Public Affairs. There is evidently a change in Health, which is good. However, it is less than Animal Care. This is an issue because people already take care of their animals before themselves. If you could deduct the amount of junk food that they consume, a decline in obesity rates would slowly take place. There is also an increase in Government and Public Affairs as once stated. However, it is a lot lower than Health and Animal Care. I understand why, however, in my opinion, it is too low. This is because people are expected to be in a certain weight bracket even though they may not have the money to afford a gym membership, which leads them to look for a park that they could possibly work out at. If they had a park more accessible to them, they could start exercising more. Finally, Animal Care is at a decent spot as of right now but not the best. Everywhere you go, you are going to see a stray animal, it is inevitable. But if you were to think about it, someone may have abandoned them for the fact that they could no longer take care of them because of their financial inability.

These issues are concerning to me because I am a future taxpayer, and I would like to assume that I would be able to benefit from the things that I am paying for. I hope that my letter has helped you realize some things that you may not have noticed before. Please consider my points when planning your budgets for years to come.

Best,
Taylor Gomez

INDEX

Pages in *italics* refer to figures and pages in **bold** refer to tables.